DESIGN TECHNIQUES FOR ORIGAMI TE...

Design Techniques for Origami Tessellations is both a collection of origami tessellations and a manual to design them.

This book begins by explaining general design methods, the history and definitions of origami tessellations, and the geometric features of flat origami, before moving on to introduce a brand-new design method: the "twist-based design method." This method generates base parts that connect "twist patterns" (that can be folded with a twist) without using a lattice. Therefore, it can generate base parts such as regular pentagons, which cannot be generated with more conventional methods, and can generate new origami tessellations connected to them.

Features:

- No proofs or formulas in the text and minimal jargon.
- Suitable for readers with a roughly middle school to high school level of mathematical background.
- Web application implementing the method described in this book is available, allowing the readers to design their own patterns.

Yohei Yamamoto is a researcher of Information and Systems at University of Tsukuba. He received his Ph.D. in engineering from the University of Tsukuba in 2023. He has been in the present post since April of the same year. His research interests center on origami design and geometric modeling techniques. The origami artworks created by him have features that are geometric flat shapes. In addition to his research activities, he is actively involved in outreach activities such as the sale of his works and the organization of workshops.

Jun Mitani is a professor of Information and Systems at University of Tsukuba. He received his Ph.D. in engineering from the University of Tokyo in 2004. He has been in the present post since April 2015. His research interests center on computer graphics, particularly geometric modeling techniques and their application to origami design. The origami artworks created by him have features that are three-dimensional shapes with smooth curved surfaces. His main books are *3D Origami Art* (2016) and *Curved-Folding Origami Design* (2019). In 2010, through an exchange with Issey Miyake, he contributed to the launch of the new 132.5 fashion brand. He also cooperated in the design of origami used in the movies "Shin Godzilla (2016)" and "Death Note Light up the NEW world (2016)". His unique origami has been well received around the world and he has received invitations to hold workshops and exhibitions in Germany, Switzerland, Italy, Israel and many other countries. His work had inspired the design of the trophy for the Player of the Match winner of each game at the Rugby World Cup 2019. He was appointed as a Japan Cultural Envoy from the Agency for Cultural Affairs and traveled to eight Asian countries to promote cultural exchanges through origami in 2019.

AK PETERS/CRC RECREATIONAL MATHEMATICS SERIES

Series Editors

Robert Fathauer
Snezana Lawrence
Jun Mitani
Colm Mulcahy
Peter Winkler
Carolyn Yackel

For more information about this series please visit: https://www.routledge.com/AK-PetersCRC-Recreational-Mathematics-Series/book-series/RECMATH?pd=published,forthcoming&pg=2&pp=12&so=pub&view=list

DESIGN TECHNIQUES FOR ORIGAMI TESSELLATIONS

Yohei Yamamoto
Jun Mitani

CRC Press
Taylor & Francis Group
Boca Raton London New York

CRC Press is an imprint of the
Taylor & Francis Group, an **informa** business

AN A K PETERS BOOK

First edition published 2024
by CRC Press
2385 NW Executive Center Drive, Suite 320, Boca Raton FL 33431

and by CRC Press
4 Park Square, Milton Park, Abingdon, Oxon, OX14 4RN

CRC Press is an imprint of Taylor & Francis Group, LLC

ISBN: 9781032453835 (hbk)
ISBN: 9781032453842 (pbk)
ISBN: 9781003376705 (ebk)

DOI: 10.1201/ 9781003376705

Typeset in Avenir LT Std
by Deanta Global Publishing Services, Chennai, India

Contents

Contents

Preface

This book presents methods to form geometric patterns by folding a sheet of paper and the works, i.e., origami, created by using the methods.

Works by folding a sheet of paper, known as "origami," have been loved by people across time and borders and continue to progress. Specifically, origami that creates tangible "shapes" such as birds, animals, and insects is popular among enthusiasts worldwide. However, the origami featured in this book does not form such specific "shapes" but rather geometric "patterns." These techniques for creating geometric patterns were spread worldwide by Shuzo Fujimoto in the late 20th century and are known as "origami tessellations." This book focuses on origami tessellations and explains the works derived from origami theories and new design methods.

Various things in our daily lives, such as walls, carpets, clothes, and pottery, are decorated with regular geometric patterns called "tessellations," enriching our lives. These patterns are rooted in traditional cultures around the world. Many of the works introduced in this book are inspired by Japanese tessellations.

Although the designs may appear complex, the theories used in their creation are quite simple. Through numerous photos and their corresponding "crease patterns," we will introduce the design methods of these works step by step, from basics to advanced.

Through this book, we can share the diverse aspects of origami, such as its artistic expression, cultural significance in tessellations, and techniques for designing based on theories.

0

Origami and Traditional Tessellation Patterns

This book introduces a new method and examples of forming tessellation patterns by folding a single sheet of paper. The geometric properties of origami have been revealed by many enthusiasts and researchers. In this chapter, we will introduce what you should know about origami as you read this book.

0.1 Background of Origami Tessellations

A tessellation is a geometric pattern consisting of shapes that fit together with no gaps and no overlaps. In particular, tessellations created by folding a sheet of paper are called origami tessellations. Figure 0.1 is one of the basic tessellations. The periodic collection of overlapping polygonal faces that are flat folded forms a geometric pattern. We can enjoy not only the different patterns on the front and back but also the shaded patterns of polygonal faces generated by a light source from behind. Various origami tessellations have been presented by many enthusiasts.

Origami tessellations are estimated to have originated in Japan. The concept originated with the *Ajisai (Hydrangea)* published by Yoshihide Momotani in Origami Nyumon (Seitosha, 1971). The work consisted of nine connected patterns of twisted folding (that are used in traditional windmills and others). In the 1980s, Shuzo Fujimoto created a series of periodic works using a similar technique, which he named "Hira-Ori" or "Twist Origami Translucent," and worked to popularize them.

In particular, Fujimoto's *Hydrangea*, which is different from the hydrangea in Momotani, is a work that still has many fans today. On the other hand, Toshikazu Kawasaki named such a series "Crystallographic Flat Origami" and explained its geometric properties from a mathematician's viewpoint.

Chris K. Palmer is largely responsible for the worldwide recognition of these works under the name origami tessellations. Influenced by Fujimoto, Palmer created works in 1994 from a single sheet of paper, consisting of a collection of three-dimensional shapes that resemble a tessellation. Following in their footsteps, many origami artists have presented origami tessellations. Representative artists include Eric Gjerde, Alexander Beber, Benjamin D. Parker, Joel Cooper, Ilan Garib, Miguel A. Gañan, and Tomoko Fuse. In addition, Robert J. Lang presents the geometry of such origami tessellations and numerous design techniques in *Twists, Tilings, and Tessellations* (CRC Press, 2018). Their works confirm that while origami tessellations are based on a simple rule, it can produce various works. We can find many works by searching for "origami tessellation" as a keyword on photo-sharing services such as Flickr and Instagram.

On the other hand, even before Momotani presented Ajisai, there were already proposals for periodic origami patterns. These patterns are known as the Yoshimura pattern (or diamond pattern), the Miura-Ori pattern, and the Ron Resch pattern. These have been used in the construction of deployable structures, such as solar panels on satellites, by combining

DOI: 10.1201/9781003376705-1

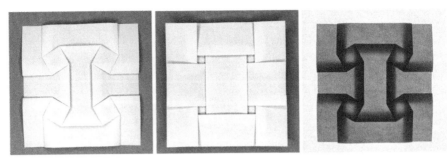

FIGURE 0.1 Example of origami tessellation (front side, back side, and shaded pattern)

small deployable modules. Herringbone pleats, which have long been used in the clothing field, can also be classified as an origami tessellation. However, these are not based on origami and are generally not considered origami tessellation origins.

In recent years, the widespread use of computers has made it possible to design patterns that are difficult to create by hand. In particular, the software "Tess," developed by Alex Bateman, is well known. We also designed many of the works presented in this book by developing our own design software. An explanation of the method to design crease patterns by connecting triangle-twist patterns implemented by this software is a major theme of this book.

0.2 Crease Patterns

The first step in learning more about origami is to observe the relationship between the shape of the folded paper and the fold lines seen when the paper is opened again. For example, if we open Figure 0.1 after folding it, a group of fold lines is visible, as shown in Figure 0.2. This collection of fold lines is called a crease pattern. There are two types of fold lines: mountain folds and valley folds. Mountain folds are lines that look like mountain ridges upon folding. In contrast, valley folds are lines that look like valleys or indentations. In this book, mountain folds are represented by solid lines and valley folds by dashed lines. In addition, for crease patterns where there is no distinction between mountain and valley fold lines, fold lines are shown as solid lines. "MV assignment" refers to the assignment

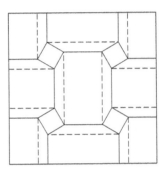

FIGURE 0.2 Crease pattern of Figure 0.1

of mountain fold or valley fold to such a fold line. Points where the fold lines intersect the inside of the paper are called interior vertices. Interior vertices where n-fold lines intersect are called n-degree vertices. The regions bounded by the fold lines or the contours of the paper are called faces. The contours of the crease patterns presented in this book are not limited to squares but include various polygons.

0.3 Basic Geometry of Flat-Foldable Crease Patterns

When creating general origami works, the process mainly involves repeating the operation of folding a paper flat. Even a crane, one of the most famous origami works, is folded flat for each step before finally spreading its wings. This operation of flat-folding paper is the basic operation of origami and is called flat folding. The fold line to be flat folded is always straight.

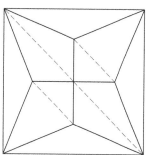

FIGURE 0.3 Bird base and its crease pattern

For example, Figure 0.3 shows what is called the bird base and its crease pattern. We can see that the crease pattern is a set of straight lines.

The origami tessellation shown in Figure 0.1 is also created by flat folding. However, the fold lines are not folded one by one in sequence. After pre-creasing the fold lines, all the fold lines are flat folded at the same time. The operation is difficult and requires practice until you get used to it. This folding technique is very important for folding the works presented in this book and will be explained in detail again in Chapter 0.7.

Much of the geometrical properties of flat folding have been clarified by earlier researchers. For example, it's known that any interior vertex of flat-foldable crease patterns satisfies the conditions described in the following three theorems.

- **Maekawa Theorem:** The numbers of mountain and valley folds at the vertex always differ by two.
- **Kawasaki Theorem:** The alternating sum of incident angles at the vertex is always zero.

- **Big Little Big Theorem:** The interior angle that is smaller than both sides is between a mountain fold and a valley fold.

Figure 0.4a shows an interior vertex contained in Figure 0.2, and Figures 0.4b and c show interior vertices contained in Figure 0.3, both of which satisfy the above three conditions. (Strictly speaking, Figure 0.4c does not have a local minimum interior angle, so it is not subject to the Big Little Big theorem.) Besides these, even if we folded a paper flatly, we can see that these three theorems are satisfied.

It should be noted that the above three conditions are necessary conditions for a crease pattern to flat fold but not sufficient conditions. For example, a crease pattern shown in Figure 0.5a, which is not subject to the Big Little Big theorem, satisfies the conditions of Maekawa and Kawasaki theorem but cannot be flat folded. On the other hand, it is known that a 4-degree vertex is always flat foldable if it satisfies the condition in Maekawa and Kawasaki theorems, and one of its smallest interior angles is between

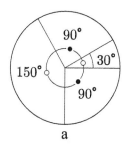
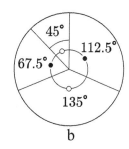
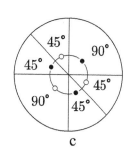

FIGURE 0.4 Interior vertices of crease patterns

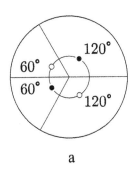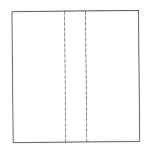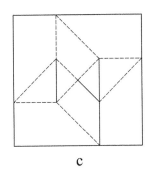

FIGURE 0.5 Unflat-foldable crease patterns

a mountain fold line and a valley fold line. We call this condition the flat-fold condition for a 4-degree vertex. (For example, Figure 0.5a cannot be flat folded because this condition is not satisfied.) A more general method of determining the flat fold at the n-th vertex is explained in detail in *Geometric Folding Algorithms* (Cambridge University Press, 2007), so refer to the book if necessary.

The crease pattern may not be flat folded due to interference between faces. To understand this situation, consider the simple crease pattern shown in Figure 0.5b: when the two valley fold lines are folded, the left-most face and the right-most face interfere with each other and cannot be flat folded. Similarly, the pattern shown in Figure 0.5c can be locally flat folded around the interior vertex, but some faces still interfere with each other and cannot be flat folded. Please note that crease patterns that can be locally flat folded may not be globally flat folded, like these examples. However, from now on, we will refer to locally flat-foldable crease patterns as "flat-foldable crease patterns" without any particular refusal.

0.4 Folded State of Crease Patterns

The operation of flat folding is closely related to reflection, which is a geometric transformation. In the flat fold, the fold line is always a straight line, and the fold is made at an angle of 180°. In other words, the face is reflected by the fold line as the axis of symmetry. For example, if we fold corner A in Figure 0.6a toward us by the dashed line, the shape A before folding and the shape after folding are mirror symmetrical. Figure 0.6b shows the paper folded again. The corner B before folding and the shape after folding are also mirror symmetrical. Opening the folded state in Figure 0.6b, we obtain the crease pattern in Figure 0.7 a.

The shape after folding a given crease pattern is obtained by determining an arbitrary

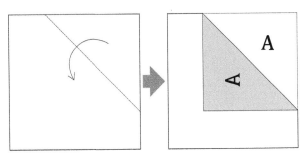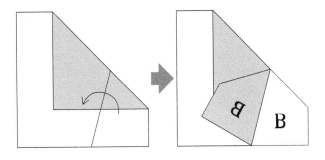

FIGURE 0.6 Operation of folding and reflection

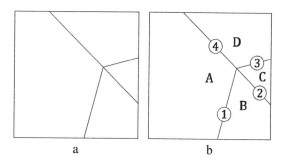

FIGURE 0.7 Crease pattern of Figure 0.6 b

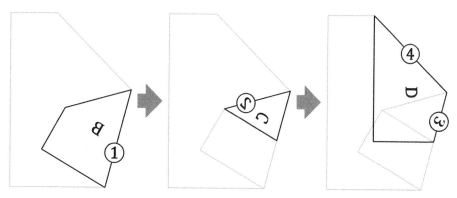

FIGURE 0.8 Method to obtain folded state

face as the seed and then recursively reflecting the faces adjacent to it, one degree at a time. The face and fold lines shown in Figure 7b are used to show the procedure in detail. As shown in Figure 0.8, first fix face A as seed and flip fold face B by fold line 1. Next, face C, which shares fold line 2 with face B, is flipped by the line. Continue by flipping face D by fold line 3. In this way, the positions of B, C, and D are determined after folding with fixed A. Although fold line 4 is not used, the folded arrangement will be such that A and D share fold line 4. (If this is not the case, the crease pattern is not flat foldable.) In this operation, there is no distinction between mountain and valley fold lines. Since the order of overlap of the faces is not considered, what we get is just a silhouette of the folded shape.

Fixing a face as a seed and folding the adjacent faces by the fold lines achieve the same result no matter what the order. For example, the pattern shown in Figure 0.8 is obtained by folding the faces in the order B, C, and D. The same result is obtained by folding the faces in the order D, C, and B. In the procedure, each time one of the faces is traversed, a reflection transformation is performed so that one of the two adjacent faces remains in its original shape and the other is a reflected shape. Therefore, if we only color the faces that are reflected after folding, the colored and non-colored faces are lined up alternately, as in Figure 0.9a. If we similarly color the crease patterns of Figures 0.2 and 0.3, they will look like Figures 0.9b and 0.9c.

0.5 Patterns for Twist Folding

Let's look at the geometrical properties of twist folding, which is used extensively in origami tessellation. As in Figure 0.10, twist folding means folding a center face (a square in this example)

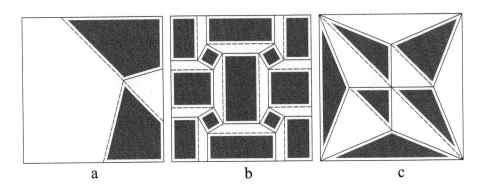

FIGURE 0.9 Colored crease patterns

while rotating the face. In this book, the center face is referred to as the rot face. Twisted folding may be prefixed by the rot face shape. For example, the folding in Figure 0.10 is called a square twist folding. In addition, crease patterns for twist folding are referred to as twist patterns. For example, Figure 0.10a is called a square twist pattern.

A twist pattern consists of a rot face and pairs of fold lines (pleats) extending outward from each side of the face. For all sides, the angles between each side and the pleat extending from the side are equal. At each vertex of the rot face, the sum of the angles facing each other is 180°, which means that the condition in Kawasaki theorem is satisfied. This means an MV assignment to satisfy the flat-fold condition for the 4-degree

vertices make this pattern flat foldable. Please note that even if locally flat foldable, crease patterns may not be globally flat folded due to interference, as already mentioned.

In a twist pattern, if the folded shape is obtained by the reflection described in the previous section using a face that is neither the rot face nor a face that makes a pleat as the seed, the rot face is rotated by twice the angle formed by the rot face sides and pleats (Figure 0.10b). Generally, twist patterns often refer to patterns with mountain folds (or valley folds) assigned to all the sides of the rot face as shown in Figure 0.10a. However, other MV assignments are acceptable in this book. Various interesting properties of twist folding will be gradually introduced in this book.

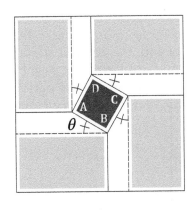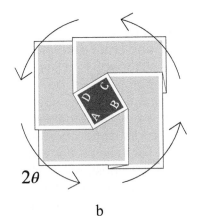

FIGURE 0.10 Example of twist folding

0.6 Tessellations

By knowing not only origami geometry but also tessellation geometry, we can expect to design more esthetically origami tessellations. This section introduces the characteristics of tessellation.

Tessellations are patterns consisting of shapes that fit together without gaps or overlaps. These patterns can be found in many aspects of our lives. We can find tessellated tiles on outdoor surfaces, walls, and indoor floors. Tessellations can also be found in clothing, wallpaper, carpets, ceramics, and everything around us. These patterns have developed in many different cultures around the world. For example, geometric patterns were used as early as 1000 B.C. in Greece. These patterns continued to develop to decorate churches and palaces. In particular, patterns found in the Islamic world, such as those shown in Figure 0.11, are known worldwide as being based on characteristic rules. The Alhambra, decorated with these patterns, has influenced many artists, including Mauritz Escher, the master of Trompe l'oeil painting. In Japan, the traditional tessellation is known as Wa-gara (that is Japanese tessellation). Typical patterns and their names are shown in Figure 0.12.

In this book, the units of shape that consist of tessellation are referred to as tiles. Tiles are limited to polygons, and tessellations are created so that adjacent tiles always touch along their entire sides (edge to edge). In the following, we introduce the geometrical characteristics of the tiles and tessellations.

First, we show the characteristics of using regular polygons as tiles. The only regular polygons that can be tiled in one type are equilateral triangles, squares, or regular hexagons, as shown in Figure 0.13; the tessellations are called regular tessellations. In these polygons, multiplying the interior angles by an integer achieves 360°. For example, the interior angle of an equilateral triangle is 60°, so multiplying by 6 gives 360°. This means that six equilateral triangles sharing a common vertex can be tiled with no gaps or no overlaps. Regular pentagons cannot be tiled in one type because multiplying its interior angle that is 108° by an integer does not equal 360°.

Tessellation can also be created by combining different regular polygon tiles. In such a tessellation, its vertices are characterized by the types of tiles that share the vertex. For example, when the tiles sharing a vertex are "equilateral triangle, equilateral triangle, square, equilateral triangle, square" in the clockwise direction, the vertex type is referred (3,3,4,3,4). Especially tessellations for which the tiles are more than one type of regular polygon and each vertex type is the same are called semi-regular tessellations. Only eight types of semi-regular tessellation exist, as shown in Figure 0.14. Figure 0.15 shows examples of tessellations for which the tiles are more than one type of regular polygon and each vertex type isn't the same.

Triangles and quadrilaterals can be tiled in one type even if they are not regular polygons, as shown in Figure 0.16. Any triangle can be copied, rotated 180°, and connected edge to edge to the original triangle to form a parallelogram.

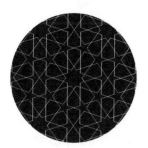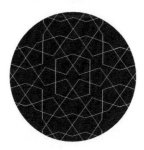

FIGURE 0.11 Examples of Islamic tessellation

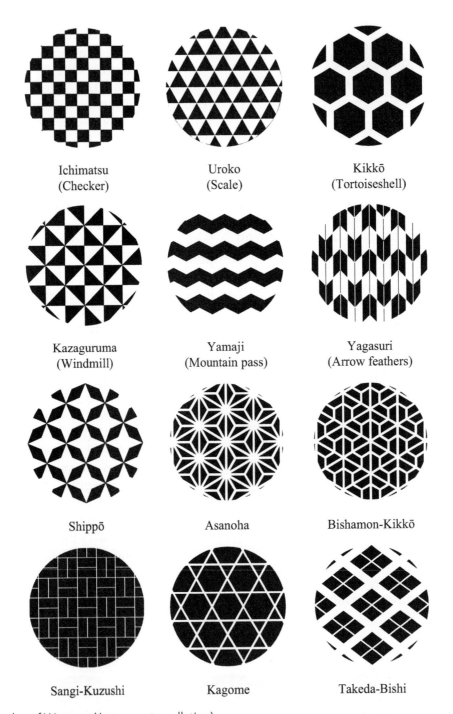

Ichimatsu (Checker)	Uroko (Scale)	Kikkō (Tortoiseshell)
Kazaguruma (Windmill)	Yamaji (Mountain pass)	Yagasuri (Arrow feathers)
Shippō	Asanoha	Bishamon-Kikkō
Sangi-Kuzushi	Kagome	Takeda-Bishi

FIGURE 0.12 Examples of Wa-gara (Japanese tessellation)

A row of tiles alternating between quadrilaterals and rotated 180°ones form zigzag bands that can be tiled. This characteristic also holds for concave quadrilaterals.

If tiles are not limited to regular polygons, the variation of tessellation is even increased. For example, regular pentagonal tiles can generate a tessellation by combining tiles matching the diamond-shaped gap shown in Figure 0.13. Figure 0.17a shows tessellations containing regular pentagons and regular dodecagons. Such tessellations have been developed in the

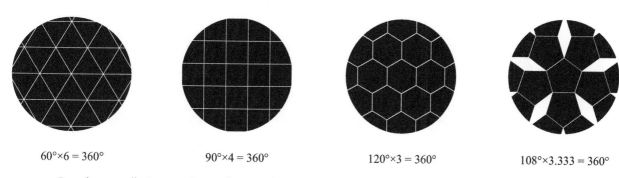

$60°×6 = 360°$　　$90°×4 = 360°$　　$120°×3 = 360°$　　$108°×3.333 = 360°$

FIGURE 0.13　Regular tessellations and tessellation of regular pentagon

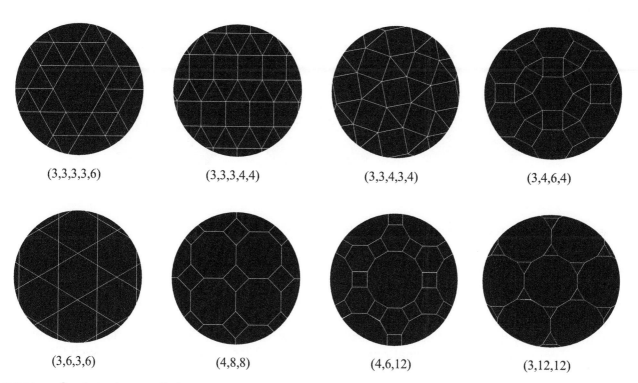

(3,3,3,3,6)　　(3,3,3,4,4)　　(3,3,4,3,4)　　(3,4,6,4)

(3,6,3,6)　　(4,8,8)　　(4,6,12)　　(3,12,12)

FIGURE 0.14　Semi-regular tessellations

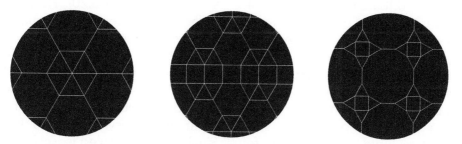

FIGURE 0.15　Examples of tessellation that aren't regular and semi-regular.

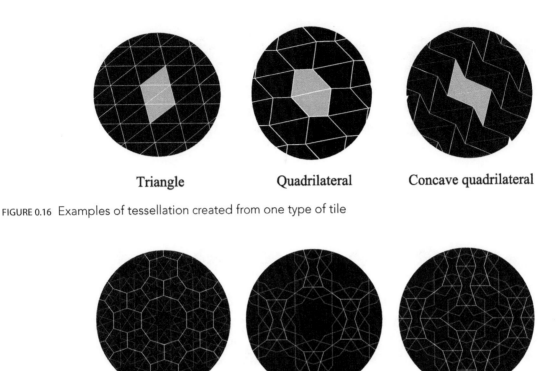

| Triangle | Quadrilateral | Concave quadrilateral |

FIGURE 0.16 Examples of tessellation created from one type of tile

| a | b | c |

FIGURE 0.17 Examples of tessellation used Islamic pattern

Islamic world. By assigning geometrical patterns to these tiles shown in Figure 0.17, the patterns shown in Figure 0.11 can be created.

Many of the tessellations already described have the property that a parallel shifted tessellation can coincide with the original one. This property is called periodic. It is known that the periodic type can be classified into 17 types called the wallpaper group. On the other hand, aperiodic tessellations are also known, as shown in Figure 0.18. Figure 0.18a is called Penrose tiling generated by two types of diamond-shaped tiles. Tessellations generated by progressively shrinking tiles, as shown in Figure 0.18b, are said to have fractal (self-similarity) properties. Figure 0.18c is a tessellation called a Voronoi diagram. It is a partition of a plane into regions close to each of a given set of points.

There are often multiple interpretations for combinations of tiles that make up a tessellation. For example, "Bishamon-Kikkō," shown in Figure 0.12, comprises one type of trapezoid.

On the other hand, the tessellation can also be regarded as consisting of one type of regular hexagonal tiles as in Figure 0.19a or two types of equilateral triangle tiles as in Figure 0.19b. That is, a set of tiles is considered a tile with a figure. Tiles in this book include such tiles.

Finally, we describe the features of tessellation, in which two tiles having figures and sharing a side are mirror symmetric. The regular tessellations of triangles or squares have mirror symmetry (Figure 0.20). However, regular tessellation of hexagons is not guaranteed symmetry since the vertices are shared by three tiles. A hexagonal tile can be regarded as a set of 12 right-triangle tiles divided by lines connecting the center to the vertices or midpoints of the sides, which can be used to create a tessellation with symmetry, as in Figure 0.21a. Similarly, right-angled isosceles triangles create such a tessellation, as in Figure 0.21b. These mirror symmetries are also utilized in this book's design of the origami tessellations.

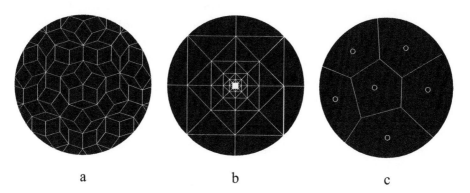

FIGURE 0.18 Examples of non-periodic tessellation

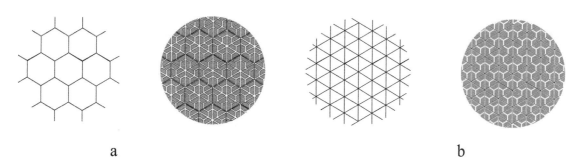

FIGURE 0.19 Example of the same tessellation with different tiles

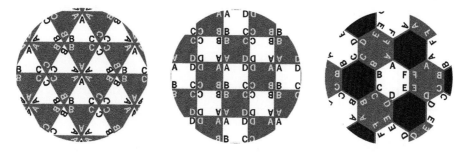

FIGURE 0.20 Regular tessellations with mirror-symmetric tiles

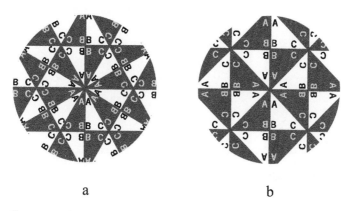

FIGURE 0.21 Example of tessellations with mirror-symmetric right-triangle tiles

0.7 Tips for Making Beautiful Folds

Finally, here we introduce tips for successfully folding origami tessellations presented in this book. Figure 0.22 shows a simple crease pattern and its folding processes. The crease pattern can be created by connecting the vertices on a 6 × 6 square grid with fold lines. As briefly mentioned in Section 0.3, to flat fold all lines at the same time, each line is pre-creased once. Since the folding operations are not performed in sequence, as in general origami, you may find it difficult. On the other hand, when you get it right, it is an exhilarating experience to flat fold the whole thing at once. In the following sections, we will introduce the steps from preparing the crease patterns to folding the piece.

Step 1. Preparation of Crease Patterns

To fold origami tessellations, you must first get a crease pattern. Some patterns in this book are based on a grid of squares or equilateral triangles, with the vertices connected by fold lines.

The grid pattern is obtained by folding a sheet of paper so that the sides are equally divided. On the other hand, many crease patterns in this book cannot be obtained by using such grid patterns. These patterns can be downloaded from our webpage at https://yohey-yamamort. github.io/book/tessellation2023/index.html. Printing the downloaded crease patterns completes the preparation.

Step 2. Creasing Fold Line on Paper

Before starting to fold printed crease patterns, score the fold line in the crease pattern using a hard pointed tool such as a ball-painted pen or a steel pen. We may not want to print crease patterns on the paper for a cleaner finish. In this case, place a printed crease pattern on a piece of paper to be folded and score the lines. Downloadable crease patterns can also be scored using a machine called a cutting plotter. A cutting plotter is a machine that automatically cuts paper by controlling the cutter blade and paper feed. It can also score fold lines by adjusting the degree of the blade edge protrusion

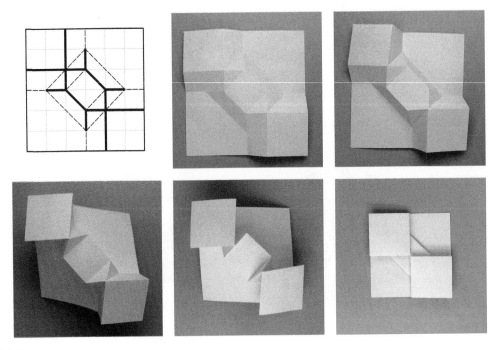

FIGURE 0.22 Folding process

and pressure. Many of the works in this book are made from crease patterns scored on Biotope paper using such a machine. Glassine paper, thin tinted paper, and Japanese paper (Wa-shi) are also recommended.

Once all the fold lines have been scored, we pinch each line to crease the lines. At the time, the mountain fold lines should be pinched from the front side of the paper. The valley fold lines should be pinched from the backside. The fold lines are added in this way to create beautiful works.

Step 3. Folding Process

Next, it is time to start the folding process. The crease pattern presented in this book requires that all fold lines be folded at the same time. First, the mountain fold lines are lightly folded into mountains, and the valley fold lines are lightly folded into valleys to form a three-dimensional shape. The interior vertex should be shaped like a peak by poking it with a sharp tool. If the paper seems to be subjected to undue force, open the paper and pre-crease the fold line again. Gradually, increase the degree of folding for all fold lines, and eventually the entire paper will flatten out. This process takes a lot of experience to get used to. This book also introduces many simple patterns, so it is a good idea to gain experience through them.

Please try to find ways that are easy to fold for you. For example, depending on the crease pattern, it may be better to flat fold the sides of the paper and then secure them with paper clips before folding the center area.

Folding operations can be difficult if papers are too small because fold lines are too fine, but if it is too large, it is, in turn, difficult to fold the center of the paper. It is best to adjust the paper to a size that allows you to control the entire fold.

Step 4. Finishing Work and Taking Photographs

Folded shapes often do not stay flat as the overlapping papers tend to open up. In such cases, lightly dampen the work with a misting spray and flatten them with a small iron to prevent them from opening. We recommend sandwiching them between sheets of clear acrylic for storage and viewing.

When you have finished your work, please take pictures of them. Geometric patterns appear not only on the front but also on the reverse side of the work. You can also capture shaded patterns by providing a light source behind the works such as by placing the works on a window pane.

These are the tips for creating beautiful works. We encourage you to experience the joy of origami tessellations. If you try again and again without giving up, you will gradually become able to do them beautifully.

1

Folding on Square Grid

In this chapter, we will introduce how to create crease patterns by placing fold lines on a square grid, as shown in Figure 1.1. The square grid can be made by dividing the sides of a square piece of paper into equal intervals with numbers such as 4, 8, and 16. We will also show how the basic fold patterns so created can be connected to each other on a square grid.

1.1. Square Twist Patterns

First, let us reintroduce square twist folding. A square twist pattern can be created by connecting the points on a 4 × 4 square grid as shown in Figure 1.1. The rot face (that is the center face) is shown in gray. Hereafter, a 4 × 4 square grid will be referred to as a "square with 4 units."

The crease pattern in Figure 1.1 can be easily made from a square piece of paper, so let's fold it. It is a good idea to draw a fold line on the paper so that you know where to fold it. Although the crease pattern is simple, you may find it difficult to fold. The same folded shape is obtained by the following procedure, which helps observe how the layers overlap. (Figure 1.2).

Step 1. Fold the paper in half, as shown in Figure 1.2a and Figure 1.2b.
Step 2. Fold as shown in Figure 1.2c and Figure 1.2d while pushing the rot face in the direction of the white arrow in Figure 1.2b.
Step 3. Pinch and fold the upper right corner of Figure 1.2d downward and pinch and fold the lower right corner except for the front one upward.

The crease pattern in Figure 1.1 can be flat folded to form a square with 2 units. The side of the folded state is folded back by 1 unit by two parallel fold lines (Figure 1.3). Since two grids are used for the folded back, the length of one side has been reduced from 4 units to 2 units. This kind of side folding will often appear in subsequent cases.

Here is an example of using a square twist pattern to create an origami tessellation. First,

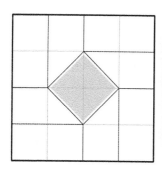

FIGURE 1.1 Square twist pattern and the folded state (front side and back side)

DOI: 10.1201/9781003376705-2

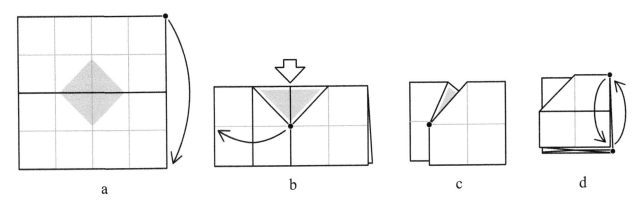

FIGURE 1.2 Process of folding square twist pattern (the gray area corresponds to the rot face)

FIGURE 1.3 Folded side

let's connect several square twist patterns. By interconnecting the patterns in Figure 1.2 and its reflected version, we can create an origami tessellation shown in Figure 1.4. Next, let's change

the MV assignment. Since the interior vertices are all of 4 degrees, the flat-fold condition for 4-degree vertices shown in Section 0.3 must still be satisfied after the assignment change. For example, when the pattern is changed, as in Figure 1.5, it becomes easier to flat fold than before the change.

Now let's change the number of grids. Figure 1.6a shows a square twist pattern created by connecting the points on a square grid

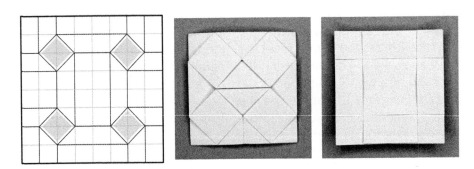

FIGURE 1.4 Connected four patterns of the twist folding

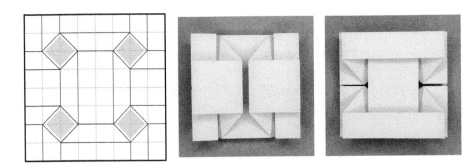

FIGURE 1.5 Crease pattern with different MV assignment of Figure 1.4

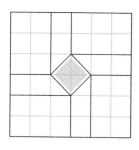
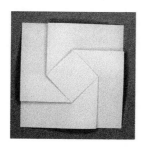

a

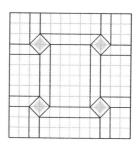

b

FIGURE 1.6 Square twist patterns changed the number of grids

with 6 units. Pleats on all four sides of the rot face are extended to reach the sides of the paper. This pattern can also be connected to create the crease pattern shown in Figure 1.6b.

We can create new crease patterns by changing the MV assignment or the paper grid. Please try to create one.

1.2. Isosceles Right Triangle Twist Pattern

Let's use a square grid to create different twist patterns. Figure 1.7 shows an example of an isosceles right triangle twist pattern placed on a square with 3 units. (The rot face is shown in gray.) Assigning mountain folds (or valley folds) to all rot face sides will prevent it from flat folding. Therefore, one valley fold and two mountain folds are assigned to the sides. When the pattern is actually folded, the rot face rotates.

It is difficult to imagine this movement from the figure, so let's try it.

Next, consider the connection of isosceles right triangle twist patterns. To connect twist patterns as tiles, the sides of the paper and the pleats should be orthogonal. Therefore, we prepare a pattern as shown in Figure 1.8. Its outline is an isosceles right triangle, and its diagonal side is 8 units, so we call it an isosceles right triangle with 8 units. The folded state forms an isosceles right triangle with 6 units. The tiles can be connected, as shown in Figure 1.9. Adjacent tiles are mirror symmetrical to each other.

Figure 1.10 shows the same twist pattern placed on an isosceles right-angled triangle tile with 6 units. Since a vertex of the rot face is touching the paper outline, a fold line of the pleat is lost in the place indicated by the black circle. The point can also be interpreted as the fold line with zero length on the tile. Even in this case, the tiles can be connected to each other like before.

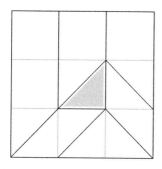

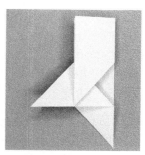

FIGURE 1.7 Isosceles right triangle twist pattern and folded state (front side and back side)

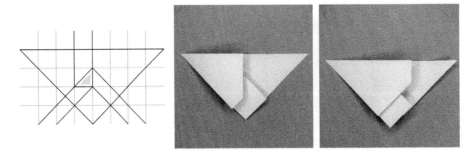

FIGURE 1.8 Isosceles right triangle twist pattern as connectable tile

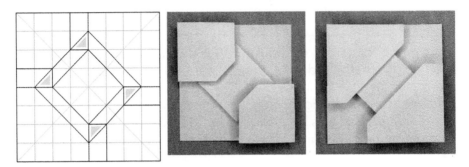

FIGURE 1.9 Connected isosceles right triangle twist patterns

FIGURE 1.10 Isosceles right triangle twist pattern with 6 units

1.3. Checker Base

This section shows patterns that look like checker patterns by folding. Checker patterns are alternating squares of two colors on a square grid, as shown in Figure 1.11a. For example, a crease pattern in Figure 1.11b forms a 2 × 2 checker pattern whose two square faces are slightly lifted by flat folding, as in Figure 1.11c. It can be created by connecting the four isosceles right triangle twist patterns in Figure 1.10.

A crease pattern in Figure 1.11c highlights the elements in Figure 1.11b. The dark gray areas are the squares that are lifted by folding. The square area with 4 units indicated by the bold line in

Figure 1.11c is referred to as a checker base. The checker base can be used as the basic unit for creating crease patterns for checker patterns. Figure 1.12 shows an example of four connected checker bases. Its folded state expresses a 3 × 3 checker pattern. The connection method to create desired checker patterns is as follows.

First, draw a desired checker pattern on a square grid, as in Figure 1.13a. The example is a 3 × 3 checker pattern consisting of squares with 3 units. The faces to be lifted up are shown in dark gray, and the others are shown in light gray. Next, rearrange the squares by inserting a space of 2 units to the interior vertices, as shown in Figure 1.13b. Assuming that a desired

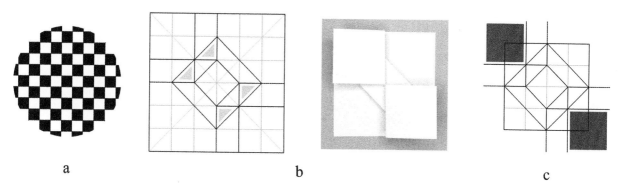

FIGURE 1.11 Checker pattern and checker base

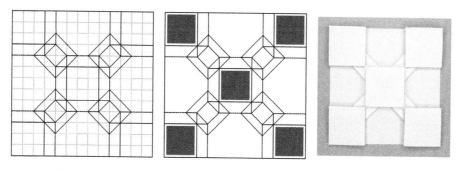

FIGURE 1.12 Crease pattern for a 3 × 3 checker pattern

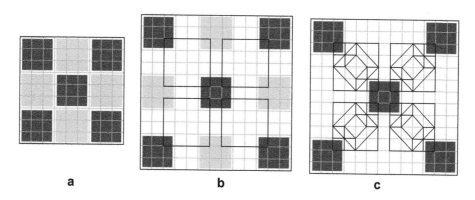

FIGURE 1.13 Procedure for drawing a crease pattern for a 3 × 3 checker pattern

checker pattern is m × m and each square is with n units, the required paper is a square with nm + 2 (m − 1) units. In this example, the paper is a square with 13 units. Next, place the checker bases in each 4 × 4 area indicated by the thick line in Figure 1.13b. Note that the orientation of the pattern differs depending on the position of the faces indicated in dark gray. By extending

pleats from the sides of each checker base pattern, the desired crease pattern is obtained, as in Figure 1.12.

Figure 1.14 shows an example of a checker pattern consisting of squares with 2 units. Try creating various checker patterns. Note that if the length of the squares is 1 unit, placed checker bases intersect each other and a crease pattern

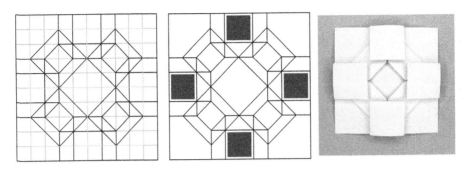

FIGURE 1.14 A 3 × 3 checker pattern consisting of squares with 2 units

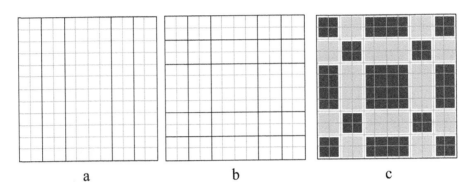

a   b   c

FIGURE 1.15 Procedure for drawing a special checker pattern

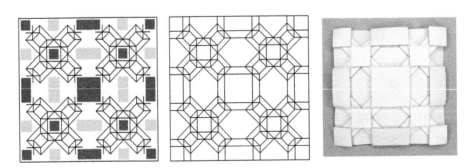

FIGURE 1.16 Crease pattern for a special checker pattern in Figure1.15

cannot be drawn. Conversely, the length of the squares can be made longer, but this increases the number of grids required and makes it more difficult to prepare them by folding.

Finally, special checker patterns are shown where the same method can be used to create crease patterns. First, draw vertical lines (Figure 1.15a), then add horizontal lines (Figure 1.15b), and finally assign two different colors, alternating, to a rectangle area (Figure 1.15c). For a checker pattern created in this way, you can also create a crease pattern, as in Figure 1.16, using the method just described. Let's use these examples to design various checker patterns.

1.4. Changing Folded Shape

As with twist patterns so far, the MV assignment of the checker base can be changed. For example, if you lift the center face of the folded shape in Figure 1.14a little and change the boundary fold lines to mountain folds, the folded shape forms a cross pattern in Figure 1.17.

Now, we take a different approach to creating new crease patterns. Look at the crease pattern in Figure 1.18a (with the MV assignments different from Figure 1.11a). The folded shape in Figure 1.18b can be transformed into the shape in Figure 1.18c by refolding two pleats in the direction of the arrows. The crease pattern of the transformed one is shown in Figure 1.18d. Further, by connecting the pattern with different MV assignments, we can create a work, as shown in Figure 1.19. Refolding pleats may not

always be possible, but once you master it, you will be able to create a greater number of works.

1.5. Crease Patterns as Connectable Tile

This section introduces a more general method of connecting tiles not limited to mirror-symmetric tiles. First, please look at the four triangle twist patterns in Figure 1.20a. These patterns are the same pattern with different MV assignments than the pattern in Figure 1.10. They are un-mirror-symmetric patterns, but connecting them creates the pattern in Figure 1.18d. Twist patterns can be connected since the connecting edge and the pleat are perpendicular, and each pleat can be shared. Note that the twist pattern contains a zero-length fold line.

This concept can also be applied to create crease patterns. First, prepare tiles of isosceles

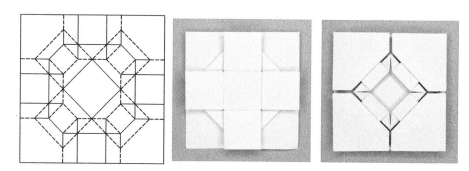

FIGURE 1.17 Crease pattern and folded state forming a cross pattern

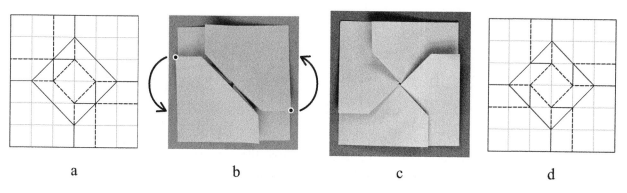

a   b   c   d

FIGURE 1.18 Refolding pleat in different direction.

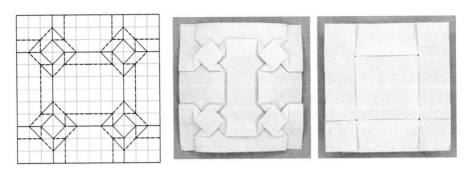

FIGURE 1.19 Work by connecting the pattern in Figure 1.18d with different MV assignment

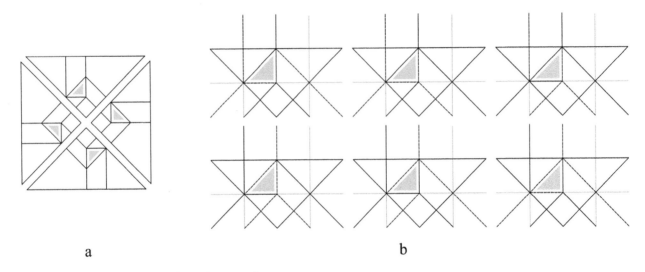

a   b

FIGURE 1.20 Connectable isosceles right triangle twist patterns

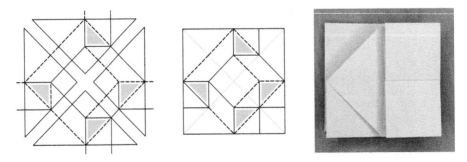

FIGURE 1.21 Example crease pattern created by connecting tiles in Figure 1.18b

right triangles in Figure 1.20b. These are isosceles right triangle twist patterns with different MV assignments. Only these six types of MV assignment make the pattern flat foldable. (The same is true for the other triangle twist patterns.)

These tiles can be connected so that the pleats are shared between two adjacent tiles, as in Figure 1.21. These tiles can also be connected, as shown in Figure 1.22. In this way, you can create a variety of works using the square grid.

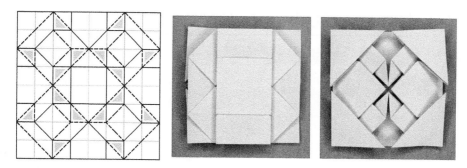

FIGURE 1.22 Crease pattern by connecting patterns in Figure 1.21

Artwork 01. Ichimatsu (Checker)

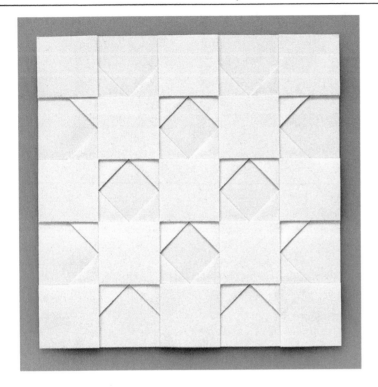

Based pattern

Back side

This is based on the checker pattern introduced several times. This pattern is called Ichimatsu in Japan. Similarly, works based on traditional Japanese patterns are denoted by their Japanese names.

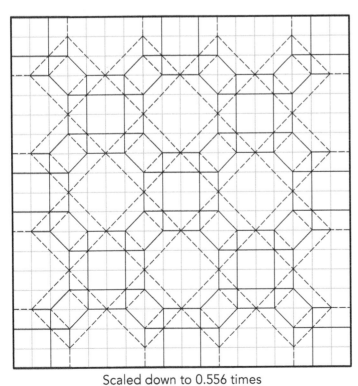

Scaled down to 0.556 times

24

Artwork 02. Rhombic tessellation

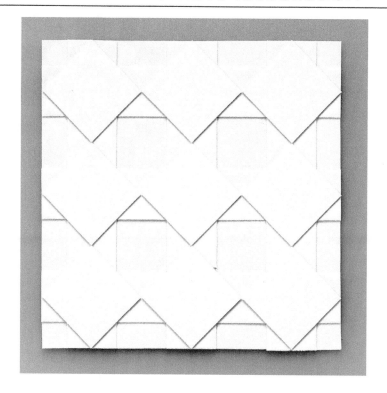

Based pattern

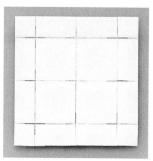

Back side

This shows a checker rotated by 45 degrees. It is only a simple operation, but it gives a very different impression.

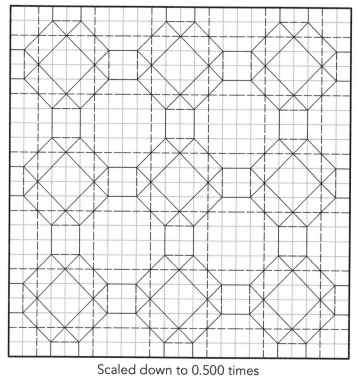

Scaled down to 0.500 times

Artwork 03. Transformed Checker

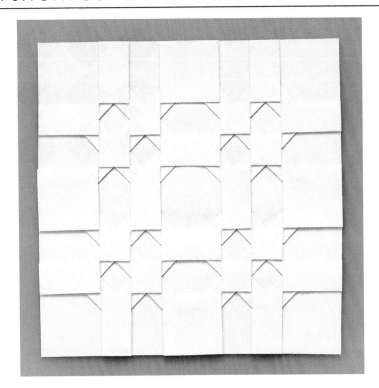

Based pattern

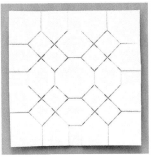

Back side

This is a checker containing rectangles designed using the method described in Figure 1.15. The crease pattern in the details can be confirmed to be checker bases.

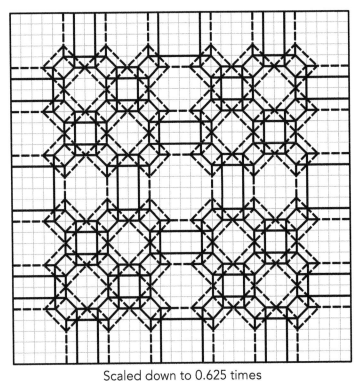

Scaled down to 0.625 times

Artwork 04. Pyramid

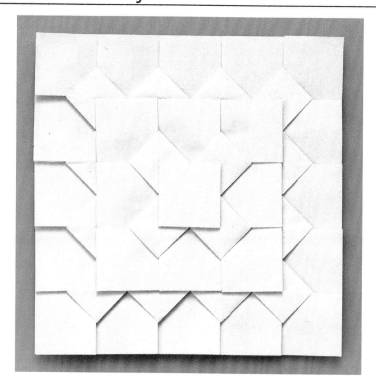

Back side

The faces of this work have pyramid-like elevation differences. In other words, the face closest to the center is stacked on top of the others. It can be made by connecting isosceles right triangle twist patterns.

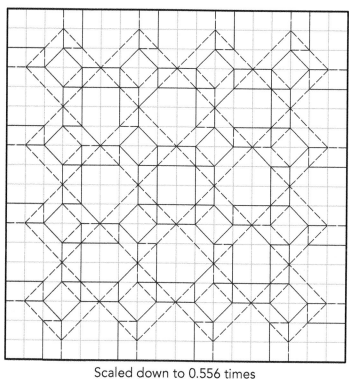

Scaled down to 0.556 times

Artwork 05. Pixel Art #1

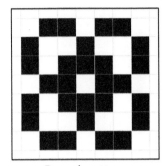

Based pattern

Back side

This work is based on two-color pixel art. As in checkers, faces corresponding to black-colored pixels are stacked above the other faces that the desired pattern is represented.

Scaled down to 0.500 times

Artwork 06. Pixel Art #2

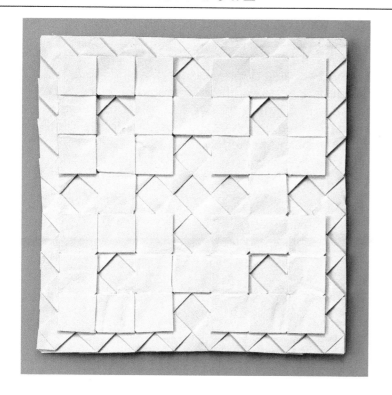

Based pattern

Back side

We can create origami works that express any two-color pixel-arts. Appendix 1 provides a design method using isosceles right triangle twist patterns for those who wish to design their own.

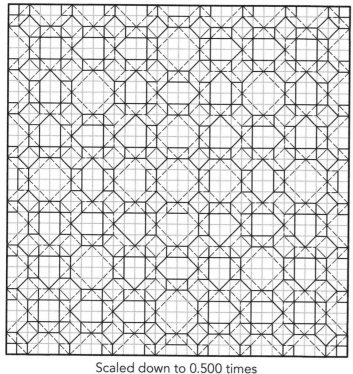

Scaled down to 0.500 times

Appendix 1: Pixel Arts Composed of Origami Tessellation

Artworks 5 and 6 express two-color pixel art. Here, we show how to design origami tessellations that express such pixel art. As in Figure 1.23, we can create works in which the faces corresponding to the gray pixel are located above other faces.

1. Place a right-angled isosceles triangle grid above a desired pixel art, as in Figure 1.23a. At this point, the center of the pixel and an 8-degree vertex of the grid should be coincided.
2. As shown in Figure 1.23b, place the pattern in Figure 1.24 in each triangular area so that the colors match. Note that tiles touching each other at an oblique side of a right-angled isosceles triangle area should be mirror symmetric to be connected.
3. The obtained crease pattern forms the desired pixel art as in Figure 1.23c by folding.

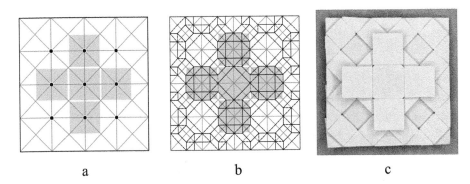

a   b   c

FIGURE 1.23 Design procedure of origami tessellations for pixel arts

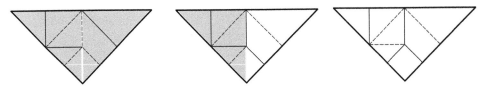

FIGURE 1.24 Three types of tiles that make up pixel arts

2

Folding on Equilateral Triangle Grid

In this chapter, we will introduce how to create crease patterns by placing fold lines on an equilateral triangle grid, as shown in Figure 2.1f. The grid can be prepared by printing a grid drawn on a computer or by dividing the sides of a hexagonal piece of paper into equal intervals. It can also be obtained by folding a square piece of paper as follows:

1. Prepare a square grid with 4 units (Figure 2.1a).

2. Fold the upper right corner with a line passing through the midpoint (white dot) of the top side so that the corner touches the leftmost vertical fold line (Figure 2.1b).

3. Fold each corner in the same way. As a result, the upper and lower sides are equally divided into 60 degrees (Figure 2.1c).

4. Exclude (or ignore) the vertical sides of a square grid (Figure 2.1d).

5. Fold the paper along lines passing through the intersection of the diagonal and

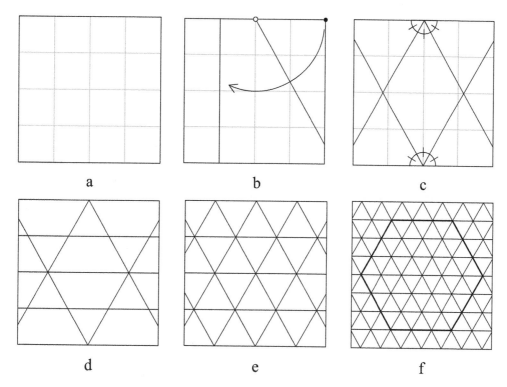

FIGURE 2.1 Process of making an equilateral triangular grid

DOI: 10.1201/9781003376705-3

horizontal fold lines so that a triangle grid is created (Figure 2.1e).

6. Further dividing the lines into equal intervals can make a finer grid (Figure 2.1f).

Hereafter, we refer to the regular hexagon shown by the bold line in Figure 2.1f as a "regular hexagon with 3 units."

2.1. Equilateral Triangle Twist Patterns

First, let's fold a single equilateral triangle twist pattern on a triangle grid. Prepare the crease pattern made by connecting the points on the regular hexagon with 2 units, as in Figure 2.2. As with the square twist pattern, folding all the fold lines at the same time forms a shape with a rotated central triangle. If this is too difficult, the same folded shape is obtained by using a pattern in Figure 2.3a. This pattern forms a three-dimensional shape in the center (Figure 2.3b). The desired shape is obtained by pressing the center, as in Figure 2.3c.

Next, let's consider the connection of the equilateral triangle twist patterns. Prepare the crease pattern as a tile made by connecting the points on the equilateral triangle with 5 units, as in Figure 2.4. Since each side of the tile and

FIGURE 2.2 Equilateral triangle twist pattern and folded state (front side and back side)

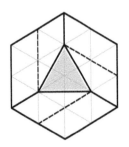

a   b   c

FIGURE 2.3 Process of folding an equilateral triangle twist pattern (the gray area corresponds to the rot face)

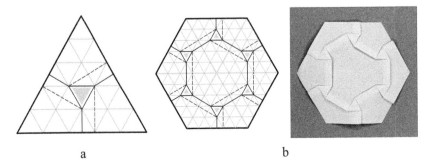

a   b

FIGURE 2.4 Six connected equilateral triangle twist patterns

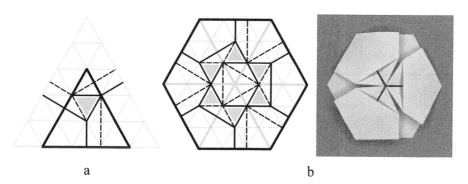

a   b

FIGURE 2.5 Changed triangle twist pattern and those of connected.

neighbor pleat is perpendicular, the tiles with mirror symmetry can be connected. Figure 2.4b shows such an example.

As with the square twist patterns, we can also change the MV assignment and the length of the pleats. An example is shown in Figure 2.5a. By connecting them, we can create a crease pattern in Figure 2.5b. In other words, these techniques are a general method to increase the variety of crease patterns.

2.2. Regular Hexagon Twist Patterns

We can also create a regular hexagonal twist pattern on a triangle grid. For example, Figure 2.6 shows a single regular hexagon pattern on a hexagon paper with 2 units. The folded state forms a shape in Figure 2.6.

The hexagon twist patterns cannot be connected as previous ones. As a test, consider the tiles in Figure 2.7a and connect them in

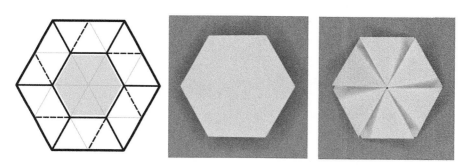

FIGURE 2.6 Regular hexagon twist pattern and the folded state (front side and back side)

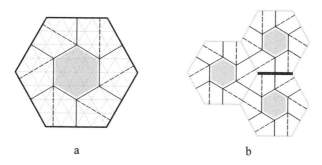

a   b

FIGURE 2.7 Example of connecting hexagon twist patterns

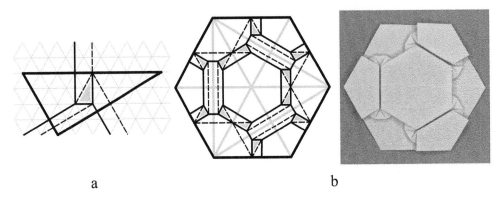

a   b

FIGURE 2.8 Right triangle twist pattern and tessellation with these tiles

Figure 2.7b. When tiles with mirror symmetry are connected, the pleats do not connect at the bold line. This is due to the fact that the vertices are shared by three tiles. Therefore, in the case of using hexagon twist patterns, they are usually combined with others.

2.3. Right Triangle Twist Patterns

As introduced in Section 0.7, using 30°–60°–90° triangle tiles, a tessellation with mirror-symmetric tiles can be generated. On a triangle grid, a triangle twist pattern as a tile can also be created with its pleats orthogonal to each side of the tile, as shown in Figure 2.8a. The tiles create a tessellation as shown in Figure 2.8b. In this way, we can create a variety of connectable triangle crease patterns on these grids.

Artwork 07. Uroko (Scale) #1

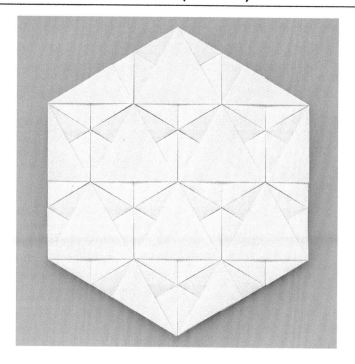

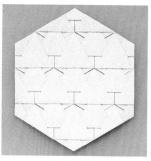

Back side

This work is based on a pattern of tiled equilateral triangles. Because of its simple crease pattern, similar works have been published by enthusiasts. If folding is difficult, practice with the base tiles.

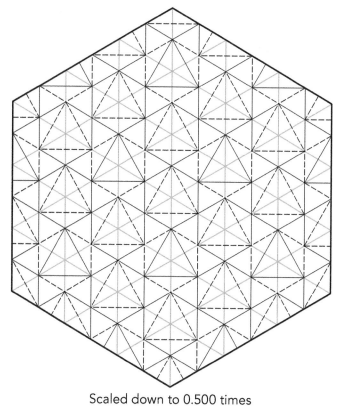

Scaled down to 0.500 times

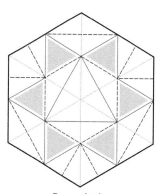

Based tile

Artwork 08. Uroko (Scale) #2

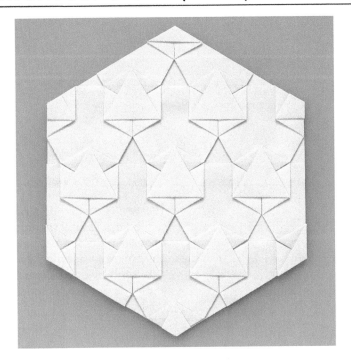

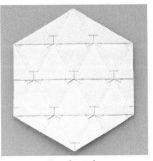

Back side

This work is based on Artwork 07, which was transformed so that there is a gap between the triangle faces. We consider it one of the most popular patterns in our work.

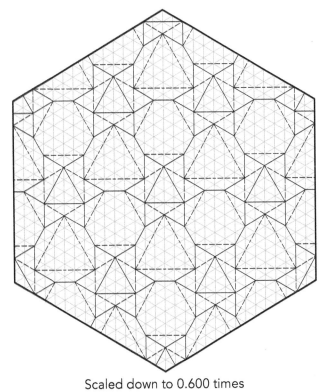

Scaled down to 0.600 times

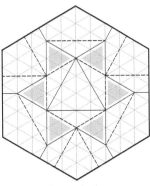

Based tile

Artwork 09. Connected Equilateral Triangles

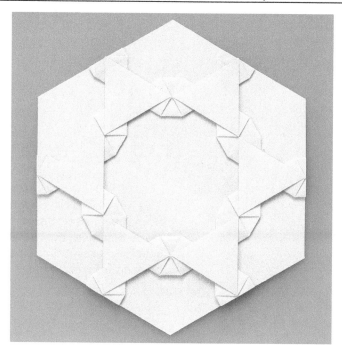

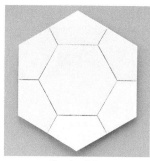

Back side

This work is made by connecting 30°–60°–90° triangle twist patterns. The base tile looks like a checker base created on an equilateral triangle grid.

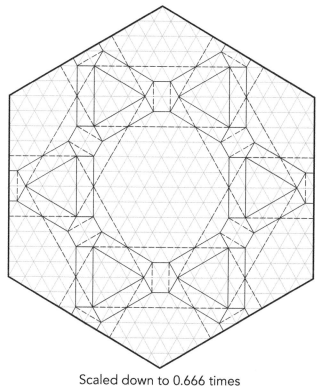

Scaled down to 0.666 times

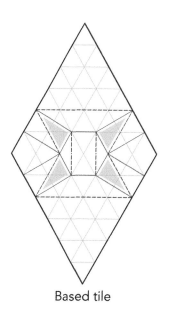

Based tile

Artwork 10. Kikkō (Tortoiseshell) #1

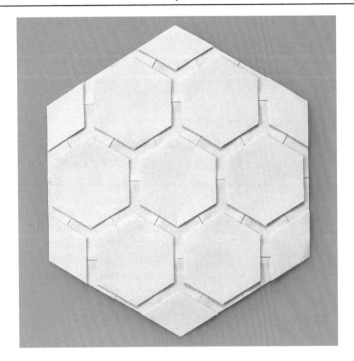

Back side

This work looks like a regular hexagonal tessellation. It can be made on a grid, but its narrow spacing makes it difficult to fold by hand from scratch.

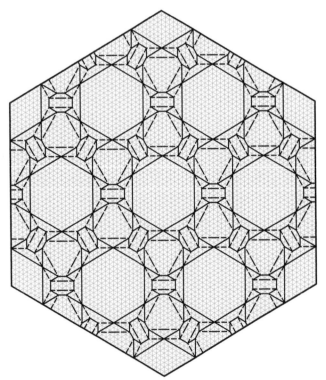

Scaled down to 0.692 times

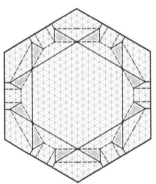

Based tile

Column 1: Grid and Twist Pattern

We have shown the methods to design various origami tessellations by connecting twist patterns on square or equilateral triangle grids. In addition to the several twist patterns introduced above, it is possible to create other twist patterns on the same grid. Examples are shown in Figure 2.9. By using these patterns, we can further increase the variety of origami tessellations.

However, the use of grids inevitably limits the angles between fold lines. In the next section, we will show how to design triangle twist patterns without using a grid.

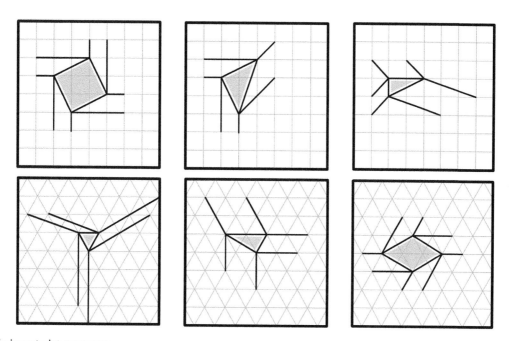

FIGURE 2.9 Various twist patterns

3

Connecting Triangle Twist Patterns

Chapters 1 and 2 introduced twist patterns created by connecting points on a square or an equilateral triangle grid and origami tessellations consisting of the twist patterns on the grid. Using this method, we can create works but only grid dependent works. This chapter shows how to create more general crease patterns of origami tessellations without using a grid.

3.1. Creating Triangle Twist Patterns

In this section, we show a method to create a single triangle twist pattern. As before, the triangle twist patterns have pleats that are orthogonal to each side of the boundary and a rot face that is similar to the shape of the boundary. The important point is that the triangle of the boundary can be freely determined.

First, create a triangle (Figure 3.1a). Next, place line segments, named "pleat bases," on each triangle's sides (Figure 3.1b). A pleat base is indicated by a black thick line. The ratio of the length of the pleat base to the length of the side corresponding to it should be the same on all three sides. Empirically, it is better to use the

ratio in the range of 0–0.25. For example, this ratio in Figure 3.1 is 0.2. Then, add perpendicular fold lines from both endpoints of all pleat bases and connect the intersections with the fold lines (Figure 3.1c). As a result, a triangle twist pattern is created. This pattern is locally flat foldable. The MV assignment is described below. In this book, the first triangle created is named "guide face," sides of guide face are named "guide sides," and the combination of a guide face and pleat bases is called a "guide."

By changing the position and length of pleat bases, we can create different triangle twist patterns. (Note the lengths of all pleat bases should be in the same ratio.) No matter how it is changed, the guide face and rot face are always similar, and the pattern is locally flat foldable. For example, Figure 3.2 shows crease patterns created by changing pleat bases from Figure 3.1. The guide in Figure 3.2a has the pleat bases whose ratio is changed to 0.1. The guide in Figure 3.2b has the pleat bases whose position is changed. The guide in Figure 3.2c has a pleat base placed outside the corresponding side. It generates a pattern in which the rot face and guide face intersect. In such a case, we change the boundary of the crease pattern to outside

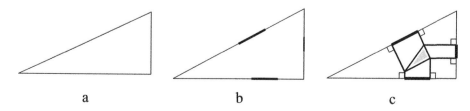

a b c

FIGURE 3.1 Method to create a triangle twist pattern

DOI: 10.1201/9781003376705-4

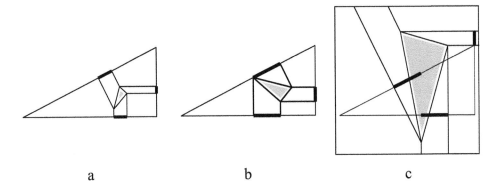

a   b   c

FIGURE 3.2 Different triangle twist patterns

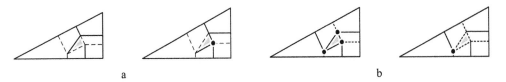

a   b

FIGURE 3.3 Examples of MV assignments of triangle twist patterns

the guide so that we can obtain a whole triangle twist pattern.

Next, we consider the MV assignment to the created pattern. Since each interior vertex is 4 degrees, it must satisfy the flat-fold condition for a 4-degree vertex described in the "Introduction" chapter. For this reason, the following procedure is recommended. First, assign one of the fold lines made up of a pleat to a mountain fold and the other to a valley fold. If both fold lines are assigned mountain folds (or valley folds), the faces are more likely to collide during the folding process as shown in Figure 0.5b. Once MV assignments of all pleats are determined, there are only two possible combinations of MV assignments for the rot face sides, based on the Maekawa theorem.

Figures 3.3a and b are examples of different MV assignments of pleats. Also, the two combinations of MV assignments of the rot face sides are shown, respectively. The left-most pattern can be flat folded. The other patterns have vertices that do not satisfy the condition (indicated by black dots).

If the rot face is created inside the guide, the boundary of the crease pattern is scaled down by folding since the ratio of the length of a boundary's side to the width of the corresponding pleat is equal on all sides. If the ratio is x, then the length of one side is 1 − 2x times by flat folding. For example, Figure 3.4a shows the crease pattern and folded state of Figure 3.3a. Since the ratio is 0.2, one side of the folded state is 0.6 times the original. The crease pattern in

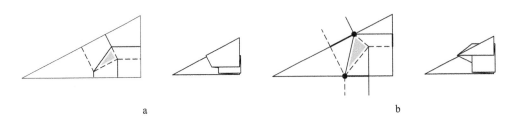

a   b

FIGURE 3.4 Similarity of the triangle twist pattern

Figure 3.4b is also scaled down when only the paper boundary is compared. This property is important when we connect twist patterns.

When a vertex of a rot face is on the boundary side, we consider that there is a zero-length fold line extending perpendicular to the side from its vertex. The vertices indicated by the black dots in Figure 3.4b are examples. In this book, MV assignments may be assumed for these zero-length fold lines as well. At that time, we interpret the virtual fold line as extending outward from the boundary as shown in the figure.

3.2. Connecting Triangle Twist Patterns

Crease patterns consisting of connected triangle twist patterns can be created with a simple rule using guides. First, a triangular tiling (each triangle can be unequal) is prepared as a set of guide faces (Figure 3.5a). Next, place a pleat base on each side. Only one pleat base is placed on one side shared by adjacent two faces. As in the case of a single guide, the ratio of the length of the pleat base to the length of the corresponding side should be the same on all sides. Finally, triangle twist patterns for each guide face are created. Triangle patterns created from adjacent guides are connected by sharing the pleat corresponding to the pleat base. Figure 3.5b shows two connected twist patterns, made from the two guides in the lower right corner. Figure 3.5c also shows six connected twist patterns. Although some of

the rot faces share a vertex, the vertices are also guaranteed flat foldable (satisfying the condition of Kawasaki theorem).

This method described so far is hereafter referred to as the "twist pattern design method." To distinguish between a single guide and a set of tiled guides, the latter is denoted Guide (whose first letter is capitalized). The shape of a Guide is defined by the polygon formed by the outline of the Guide. For example, the shape of the Guide in Figure 3.5a is a square. The sides of a shape are called Guide sides. It may consist of multiple guide sides. When all rot faces are contained inside a Guide, the boundary of the Guide is scaled down by folding. MV assignments are recommended to be made in the order of pleats and sides of rot faces, as with a single triangle twist pattern. Figure 3.6 shows a crease pattern of Figure 3.5c with MV assignment and the folded state. (Again, the "flat-foldable" referred to in this book is only guaranteed locally flat foldable around each vertex.)

The twist pattern design method is general and applicable since it uses no square or equilateral triangle grid. The crease patterns introduced in the previous section can also be constructed using this method. In other words, once you have mastered the twist pattern design method, you can create crease patterns created by different methods using only this method. For example, a guide with the right-angled isosceles triangle face in Figure 3.7a generates the triangle twist pattern shown in Figure 1.20.

Next, we introduce peculiar crease patterns that can be created using the twist pattern

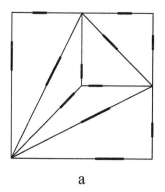

a

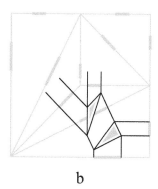

b

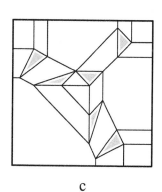

c

FIGURE 3.5 Twist pattern design method

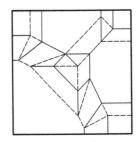
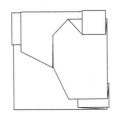

FIGURE 3.6 A crease pattern of Figure 3.5c with MV assignment and the folded state

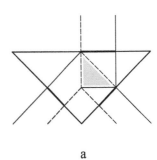
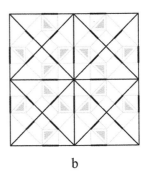

a   b

FIGURE 3.7 Guides for existing crease patterns (a: Figure 1.20, b: Figure 1.22)

design method but are difficult to consider as a set of connected triangle twist patterns.

A Guide shown in Figure 3.8a is shaped as a triangle consisting of two triangle guides. The two pleat bases on the base of the boundary touch each other at their endpoints. These pleat bases generate overlapped fold lines, as shown in Figure 3.8b (indicated by the bold lines).

Interestingly, crease patterns with overlapped fold lines removed are also flat foldable. In the case of this (Figure 3.8c) it can be interpreted as two pleats combined into one.

Figure 3.9 is an example where two rot face sides overlap each other. The Guide is a square consisting of two triangle guides. One of the guides generates a triangle twist pattern in Figure 3.9b. When the other guide generates a triangle twist pattern, the sides of the two rot faces overlap in Figure 3.9c. The crease pattern from which the overlapped fold lines are removed, shown in Figure 3.9d, is a square twist pattern, so it is clearly flat foldable.

Since overlapped fold lines can be removed, general twist patterns can be interpreted as a set of triangle twist patterns. For example, Figure 3.10a shows a regular hexagonal twist pattern consisting of six triangle twist patterns. Figure 3.10b shows a square twist pattern consisting of three triangle twist patterns. In this example, the fold lines of the pleats are also

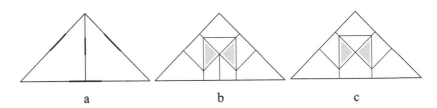

a   b   c

FIGURE 3.8 Overlapped fold lines of pleats and removal

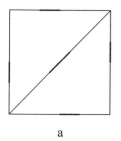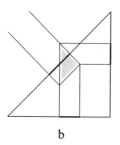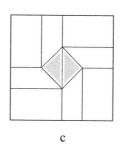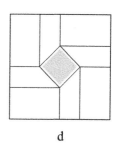

a   b   c   d

FIGURE 3.9 Overlapped fold lines of rot faces and removal

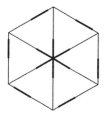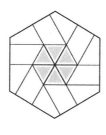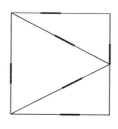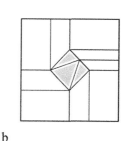

a   b

FIGURE 3.10 Typical twist patterns and Guides to generate them

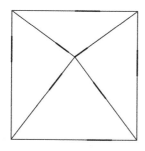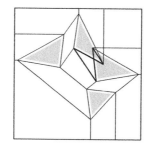

FIGURE 3.11 Intersected fold lines

overlapped. The removed fold lines of such twist patterns are obtained by the procedure of dividing the rot face into an arbitrary set of triangles and reconstructing twist patterns with each triangular face as the rot face. The pleats can also be easily reconstructed since the angles between each side and the pleat extending from the side are equal for all pleats. Depending on the division, the same twist pattern can be interpreted as different sets of triangle twist patterns, as in Figures 3.9c and 3.10b.

When fold lines intersect each other, as shown in Figure 3.11, the design is considered to have failed. In this case, the position of pleat bases should be changed. Since it is difficult to perform this trial-and-error process manually, the authors have developed software to assist in this process. The software is available on the web so that you can use it. For details, please refer to Column 02.

3.3. Design for Regular Polygon Patterns

In this section, we introduce how to create a regular polygon pattern with rotational and mirror symmetry using the twist pattern design method. First, divide a regular n-gon into 2n similar right triangles by line segments

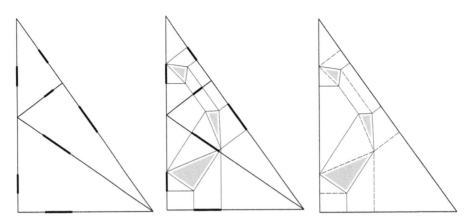

FIGURE 3.12 Triangle crease pattern with interior angles of 36°, 54°, and 90°

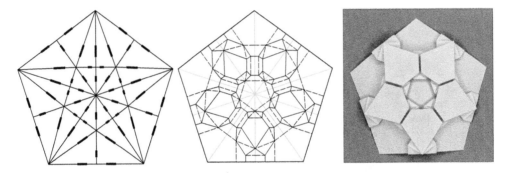

FIGURE 3.13 Origami tessellation with regular pentagonal boundary

connecting the center and the vertices and by line segments connecting the center and the midpoints of the sides. For example, a regular pentagon yields 10 triangles with interior angles of 36°, 54°, and 90°. Next, triangulate the triangle further. Obtain a guide by interpreting the set of triangles as a set of guide faces and create a crease pattern whose rot faces are located inside the boundary. Figure 3.12 shows an example. Tiling the crease patterns so that they are mirror symmetric on shared sides can create a crease pattern with the original regular polygon as the boundary. Tiling the crease patterns in Figure 3.12 yields a crease pattern with a regular pentagonal boundary in Figure 3.13. This example of a regular pentagon cannot be designed

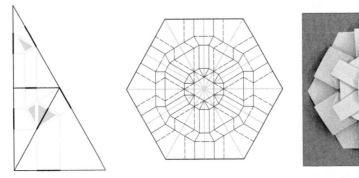

FIGURE 3.14 Origami tessellation with regular hexagonal boundary

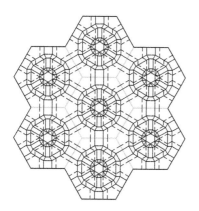

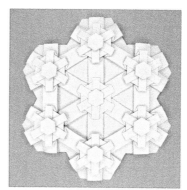

FIGURE 3.15 Connecting crease patterns in Figure 3.14

using a regular grid, showing the superiority of twist pattern design method.

If the regular polygonal pattern created is a regular triangle, square, or hexagon, it can be tiled further. For example, tiling regular hexagon crease patterns shown in Figure 3.14 yields a crease pattern shown in Figure 3.15. Thus, combining simple triangle twist patterns can generate a complex origami tessellation.

Artwork 11. Star #1

Back side

This work is a star made by folding a regular pentagonal piece of paper. It was designed by changing the positions of the pleat bases in Figure 3.12.

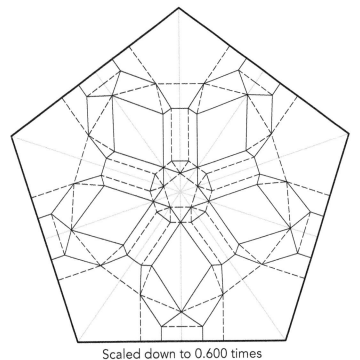

Scaled down to 0.600 times

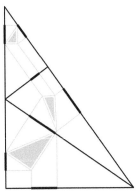

Guide for based tile

Artwork 12. Star #2

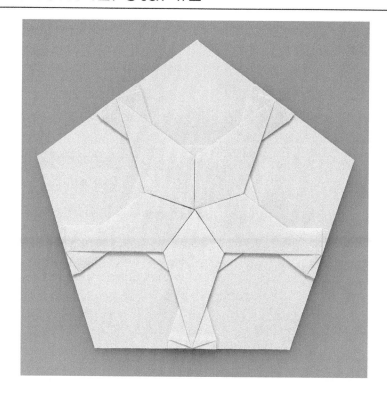

Back side

This work is Artwork 12 modified to represent a sharper star. The twist pattern design method makes it easy to create such works in which only the angle of the fold line differs.

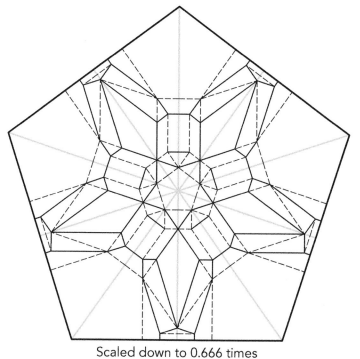

Scaled down to 0.666 times

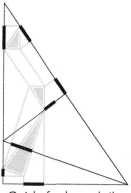

Guide for based tile

Artwork 13. Snowflake #1

Back side

This Guide for based tiles is also similar to the one of Artwork 11. You can see how the basic tile can be applied to various regular polygons by modifying it.

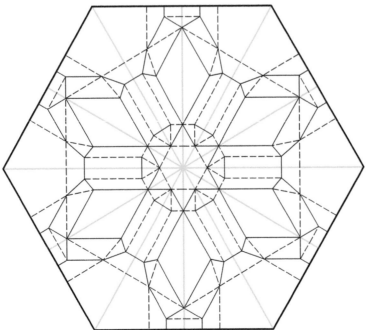

Scaled down to 0.600 times

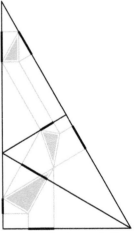

Guide for based tile

Artwork 14. Square Kazaguruma (Windmill)

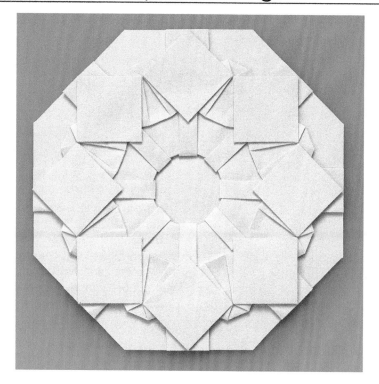

Back side

This crease pattern has some narrow spacing between the fold lines. To create beautiful works, it is sometimes necessary to construct difficult fold lines like this.

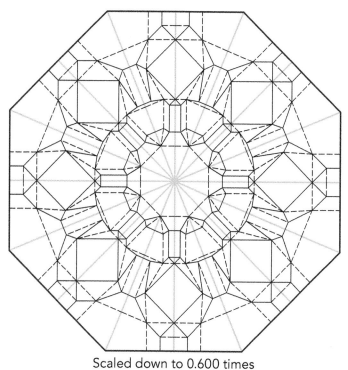

Scaled down to 0.600 times

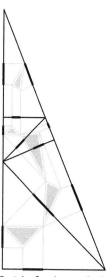

Guide for based tile

Artwork 15. Islamic Regular Decagon

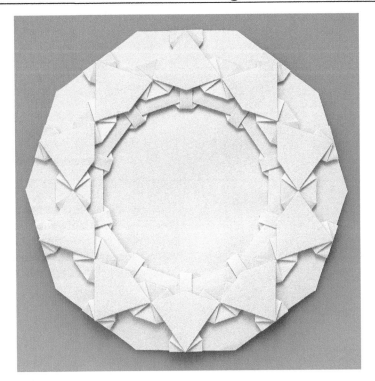

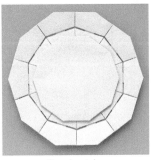

Back side

This is part of the Islamic pattern introduced in the "Introduction" chapter. We will show in the next chapter these parts can be tiled to create more complex Islamic patterns.

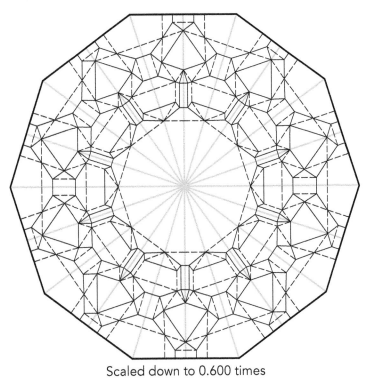

Scaled down to 0.600 times

Guide for based tile

Artwork 16. Kikkō (Tortoiseshell) #2

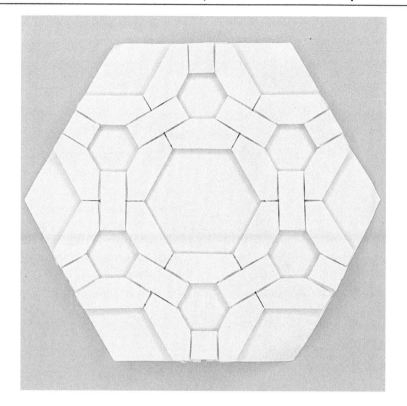

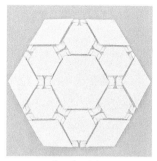

Back side

I tried to hide other faces behind the faces that express the hexagons. It looks like a simple pattern, but it is very hard to fold.

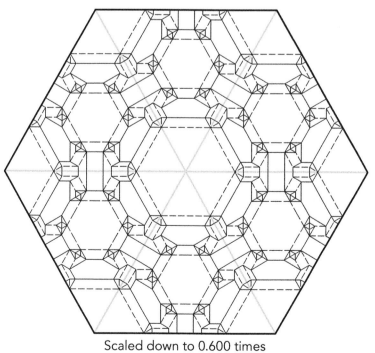

Scaled down to 0.600 times

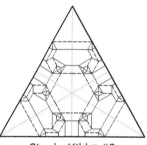

Single Kikkō #2

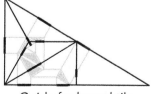

Guide for based tile

Artwork 17. Tiled Snowflake #2

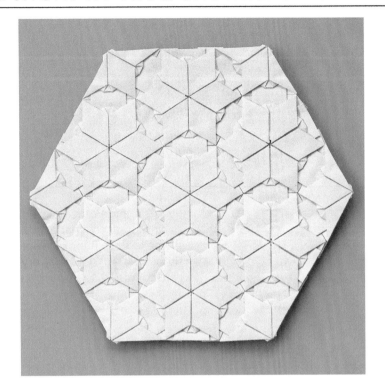

Back side

This work is tiled with a similar pattern to the one in Artwork 13. The pattern is redesigned to change its orientation relative to the boundary.

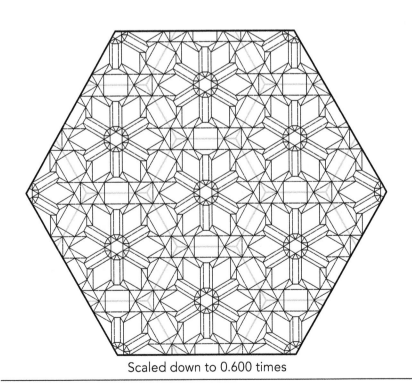

Scaled down to 0.600 times

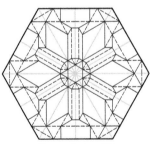

Snowflake #2

Guide for based tile

Artwork 18. Shippō

Back side

Shippō is one of our favorite patterns. The crease pattern has many places where different fold lines are aligned on the same line, so it is easier to fold than it looks.

Scaled down to 0.545 times

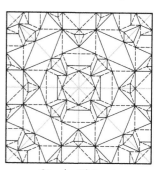

Single Shippō

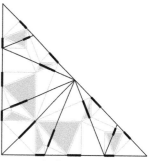

Guide for based tile

Column 02: How to use Triangle Twist Pattern Maker

The twist pattern design method is a versatile method that can be used to create crease patterns by connecting triangle twist patterns. However, the position and the length of pleat bases must be changed to prevent a crossing of fold lines. Since it is difficult to perform this trial-and-error process manually, we have implemented software named "Triangle Twist Pattern Maker." The software generates crease patterns interactively based on the input Guides. The software runs on a Web browser, so you can use it right after launching the web page.[1] Since the functions and interface of the software may be updated in the future, this column only introduces its basic usage of it.

First, when the software is launched, a triangle appears on the screen (Figure 3.16a). Based on this triangle, prepare a triangle tiling as a set of guide faces. By clicking on any point, a triangle connecting the point and the nearest side's endpoints can be created. Vertices can be moved by mouse drag. If a vertex is moved to make a triangle with zero areas, the vertex is deleted (Figure 3.16b).

Next, operate pleat bases. By clicking on the "Crease mode" tab on the right side of the screen, the system will switch to the mode of pleat base operation (Figure 3.16c). (Click on the "Polygon mode" tab to switch to operate triangles.) The screen shows pleat bases indicated by a bold line and a crease pattern generated by the Guide. The position of the pleat base can be moved by mouse drag. Since the length of the pleat base is proportional to the corresponding guide side, it can be changed by manipulating the "pleat width rate" slider bar on the right side of the screen. While these operations are being performed, the crease pattern is updated in real time.

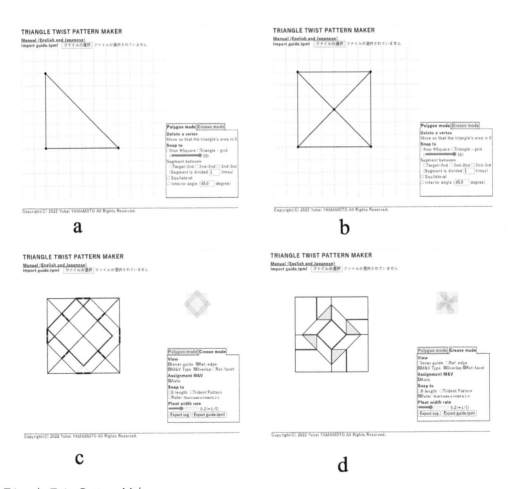

FIGURE 3.16 Triangle Twist Pattern Maker

Finally, process the MV assignment. On the screen, mountain folds are indicated by red lines and valley folds by blue lines. In this system, pleats are assumed to consist of one mountain fold line and one valley fold line. Clicking on a pleat switches this MV assignment. Each side of a rot face is automatically assigned a mountain or valley fold, considering the surrounding pleats and the flat-fold condition for a 4-degree vertex. If there is no MV assignment to satisfy the conditions, the fold line is grayed out.

With these operations, the desired crease pattern can be created (Figure 3.16d). Once the crease pattern is completed, try folding it. To do so, you can print out the screen image or you can use the SVG file saved by clicking the "Export as SVG" button on the right side of the screen.

The operations described here are basic. Other features to simplify designing include the ability to snap to triangle vertices and pleat bases and a button to show/hide each element. More features may be added in subsequent updates.

Appendix 2: Changing Length of Pleat Base

The Triangle Twist Pattern Maker allows the user to change the length of all pleat bases. To implement this operation, the following rules are defined. Figure 3.17a illustrates a guide side and the pleat base. s is the distance from one endpoint of the guide side to the endpoint of the pleat base. t is the distance from the other endpoint of the guide base to the other endpoint of the pleat base. Our system assumes that the ratio s:t is maintained even if the length is changed. For example, Figure 3.17c shows the length of the pleat base in Figure 3.17a doubled. In this case, the distances u and v between the endpoints must satisfy s:t = u:v. Similarly, Figure 3.17d shows the length of the pleat base in Figure 3.17b doubled.

A point obtained by setting the length to 0 is the origin point of the pleat base, since the point is independent of the length. If a pleat base is on the guide side, the point is an internal division point of the side. If a pleat base is outside the guide side, as shown in Figure 3.17b, then the point is an external division point of the side.

We will show in the subsequent appendices that deformations based on this rule yield even more interesting results.

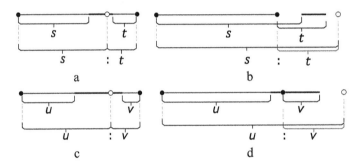

FIGURE 3.17 Pleat bases and their origin point

Note

1 "TRIANGLE TWIST PATTERN MAKER." http://www.cgg.cs.tsukuba.ac.jp/~yamamoto_y/orihole/application/ttp_maker.html (accessed on April 26, 2023)

4

Connecting of Different Crease Patterns

Origami tessellations, introduced in Chapter 3, were a concatenation of a single type of crease pattern as a tile. However, different crease patterns can be connected if they have sides that fold to the same shape. This chapter shows how to design crease patterns that can be connected in this way.

4.1. Connectable Side of Boundary

In the twist pattern design method, if all rot faces are located inside the boundary of Guide, the boundary is scaled down by flat folding. If two such Guides are connected so that they share a single Guide side and its pleat bases, the generated crease patterns are also connected. The patterns consisting of mirror-symmetric patterns in Chapter 3 were created based on this feature. Origami tessellations can be created by pre-designing multiple crease patterns that can be connected and then connecting each other. The pre-designed patterns can be regarded as "tiles" because they are the units of the tessellation.

Here are some concrete examples. Figure 4.1 shows two types of connectable square tiles. The connectable Guide side, the pleat bases, and the guide vertices are highlighted. It also depicts shared pleats extending outside the boundaries. When the sides are touched, the pleat bases that achieve the same pleat also coincide, and they connect so that the length of the pleats extending outward is zero. In the following sections, we show how to design these connectable and different tiles.

4.2. Regular Tessellations

First, focus on square crease patterns. A square crease pattern with an MV assignment in Figure 4.2 has a mirror and rotational symmetry, designed using the method described in Section 3.3. This pattern can be tiled so that it creates a regular tessellation. By using the Guide for the pattern as a basis, different and connectable crease patterns can be designed. To do so, move any of the pleat bases that are not on the boundary. Figure 4.3 shows two example patterns designed by the method.

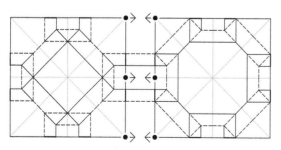
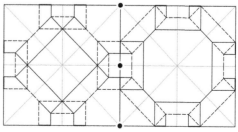

FIGURE 4.1 Connection of square tiles

DOI: 10.1201/9781003376705-5

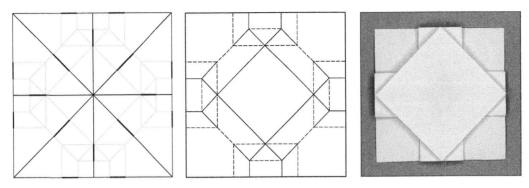

FIGURE 4.2 Example square crease pattern as tile

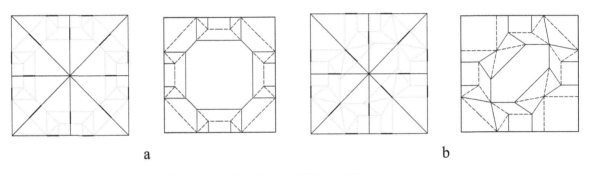

a   b

FIGURE 4.3 Example crease patterns by moving pleat bases of Figure 4.2

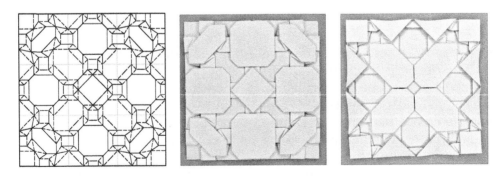

FIGURE 4.4 Regular tessellations by different square crease patterns

Since the MV assignments to the pleat to fold the side are not changed, they can be tiled in any combination. Figure 4.4 is an example of tessellation. Figure 4.3b does not have a four-fold rotational symmetry but can still be connected in any orientation.

Next, let's change the set of guide faces of the basic Guide while the boundary remains the same. Since the boundary does not change, the resulting crease patterns can be connected to each other. Examples are shown in Figure 4.5. These patterns can also be tiled as in Figure 4.6.

Similarly, we can design several connectable equilateral triangular or regular hexagonal crease patterns. Figures 4.7~ show such examples. Complex origami tessellations can be created by creating many tiles in these ways.

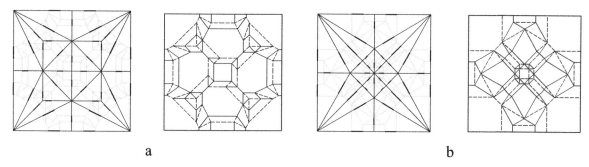

FIGURE 4.5 Example crease patterns by changing the guide faces of Figure 4.2

FIGURE 4.6 Regular tessellations by using tiles in Figure 4.5

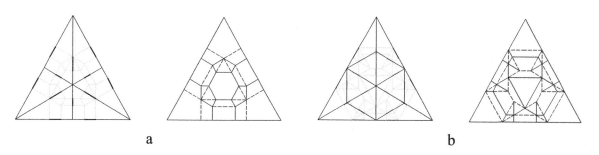

FIGURE 4.7 Example equilateral triangular crease patterns

4.3. Tessellation with Equilateral Polygons

Many tessellations, such as semi-regular tessellations, are sets of tiled equilateral polygons. This section shows how to design similarly connectable crease patterns with equilateral polygonal boundaries.

As a matter of fact, the regular triangle, square, and hexagonal crease patterns created in Section 4.2 are examples of such connectable

ones. It is possible to create origami tessellations by tiling those patterns like a semi-regular tessellation, as shown in Figure 4.11.

To create connectable crease patterns, it is first necessary to determine a shared Guide side and pleat bases on the side. The Guide side can be mirror symmetric to the midpoint to connect in any orientation. Figure 4.12a shows an example of a Guide side composed of two guide sides. This is also common to the Guides introduced in Section 4.2. Equilateral polygons

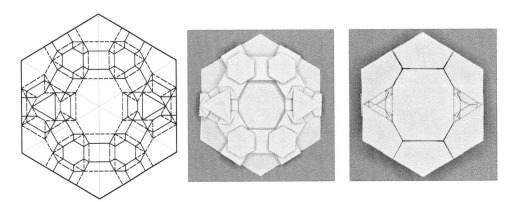

FIGURE 4.8 Regular tessellations by using tiles in Figure 4.7

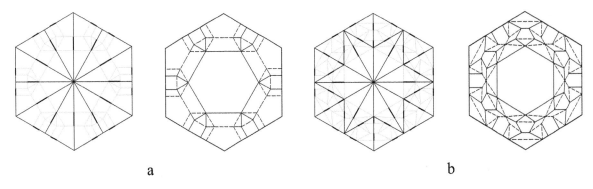

a   b

FIGURE 4.9 Example hexagonal crease patterns

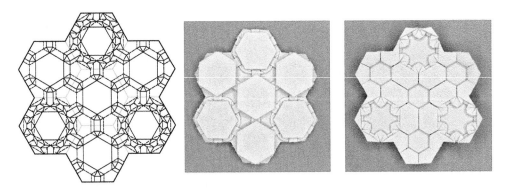

FIGURE 4.10 Regular tessellations by using tiles in Figure 4.9

that combine the Guide side are a boundary of a Guide (Figure 4.12b). Then, by properly filling in the inside of the Guide, connectable crease patterns can be created.

There is one important point to note here. In this method, a boundary shape is determined first. Depending on the shape, rot faces may always intersect the boundary no matter how the inner fold line is created. In other words, unconnectable sides are yielded. Remember that this is a special case, but it can happen. More details are introduced in Appendix 3.

Boundary shapes that can be created in this way are not limited to the same ones in

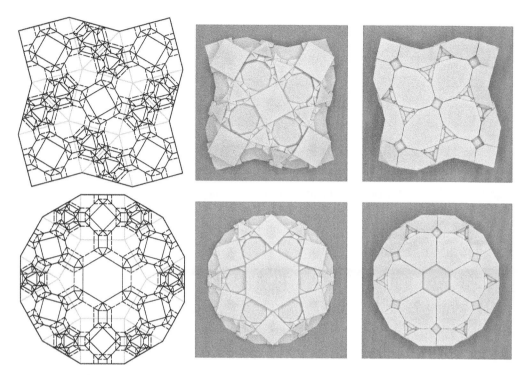

FIGURE 4.11 Origami tessellations like semi-regular tessellations (3,3,4,3,4) and (3,4,6,4)

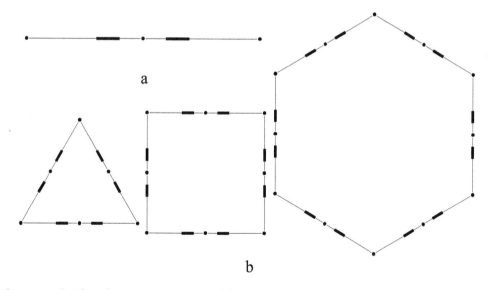

FIGURE 4.12 a: Common Guide side to create connectable equilateral crease pattern, b: examples of Guide boundary

Figure 4.12. For example, Figure 4.13 shows an origami tessellation with tiled pentagonal and rhombic patterns. Their tiles are created by using the Guide side in Figure 4.12a as a constraint (Figure 4.14). This way, a new origami

tessellation can be created based on your favorite tessellation patterns.

Finally, a special case is described. Section 3.2 shows two pleats with overlapped fold lines can be interpreted as a single pleat. Considering

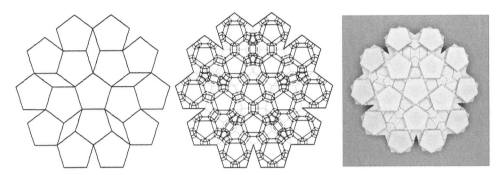

FIGURE 4.13 Origami tessellation tiled pentagonal and rhombic crease patterns

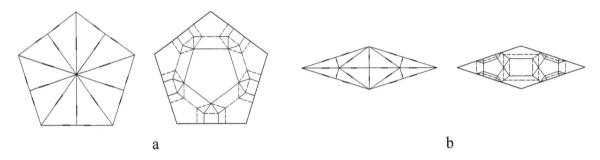

a   b

FIGURE 4.14 Crease patterns comprising the tessellation in Figure 4.13

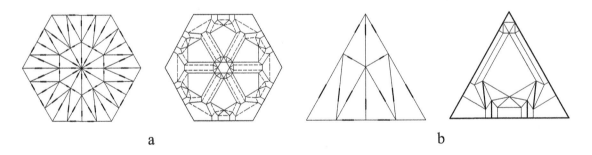

a   b

FIGURE 4.15 Crease pattern with overlapped pleats to fold boundary

this, even Guide sides with different numbers of guide sides can be connectable. Figure 4.15a shows a regular hexagon pattern and the Guide (the 1/6 sector of the pattern and the Guide are shown in Figure 4.15b). Four pleat bases are located on each Guide side. Since the overlapped fold lines (shown in bold line in Figure 4.15b) can be removed, it appears that two pleat bases are actually located on each Guide side. Then, since some of the pleat bases touch each other at their endpoint, this Guide side can be interpreted as the same as Figure 4.12a, and

the crease pattern in Figure 4.14a can also be connected to the others.

4.4. Combining Crease Patterns Having Different Guide Sides

So far, we have described methods to design origami tessellations by connecting crease patterns having common Guide sides as tiles.

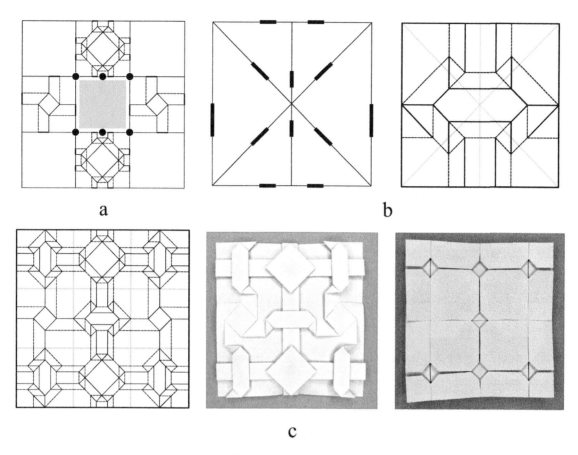

FIGURE 4.16 Combining crease patterns having different Guide sides

On the other hand, we can combine crease patterns having different Guide sides to create a work. These patterns cannot be directly connected but may be connected via another crease pattern. For example, we show a method to design an origami tessellation that includes the crease patterns shown in Figures 3.09 and 4.2. Figure 4.16a shows those patterns located around a square area. To create the desired tessellation, we need to design a crease pattern that locates in the area and connects with the surrounding patterns. The Guide for the area has a boundary given as a constraint from the surroundings Guides. Figure 4.16b is an example of a Guide satisfying the constraint. Using the pattern obtained from the Guide, we can create the desired pattern in Figure 4.16c. This approach is useful for designing complex origami tessellations. Note that crease patterns that can be connected in the approach are limited to those with the same ratio of the length of the guide side to the length of the pleat base.

Artwork 19. Ninja Star

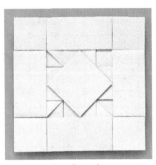

Back side

This work represents one of the most famous origami works, Ninja Star. Since Ninja Star is rotationally symmetric, the base tiles were designed to be connected on any side.

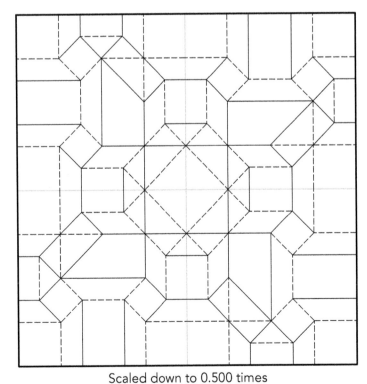

Scaled down to 0.500 times

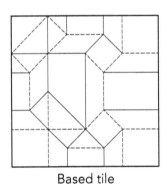

Based tile

Artwork 20. Metamorphose of Snowflakes

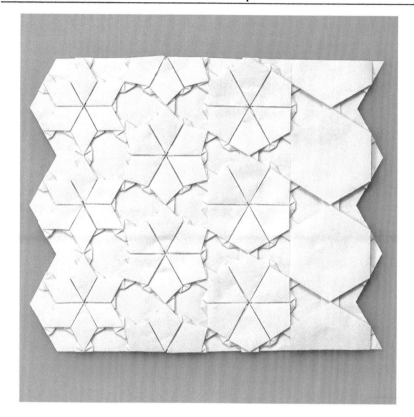

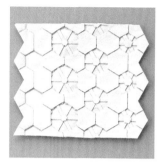

Back side

This work is tiled with four types of pattern: Artwork 13, Figure 4.9a, Figure 4.15a, and the below tile. As a result, a gradually changing pattern could be expressed.

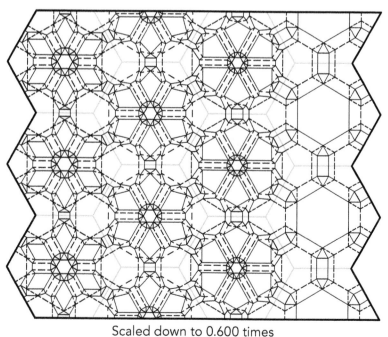

Scaled down to 0.600 times

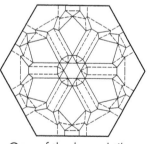

One of the based tiles

Artwork 21. Tiled Flowers

Back side

This work follows a semi-regular tessellation of (3,4,6,4). Three types of tiles are used: Artwork 13, Figure 4.5b, and Figure 4.7b.

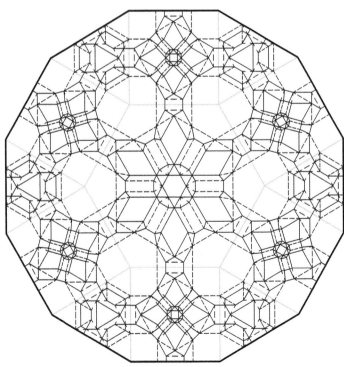

Scaled down to 0.600 times

Artwork 22. Tilled Square Kazaguruma

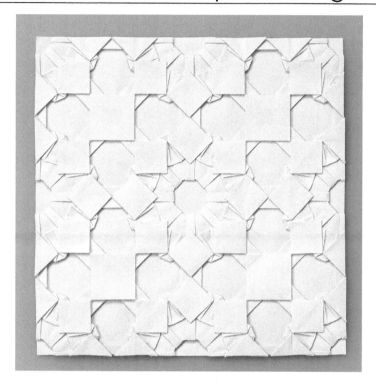

Back side

This work follows a semi-regular tessellation of (4,8,8). Two types of tiles are used: Artwork 14 and Figure 4.2 It looks like a variant of the checker pattern.

Scaled down to 0.600 times

Artwork 23. Tiled Regular Pentagons

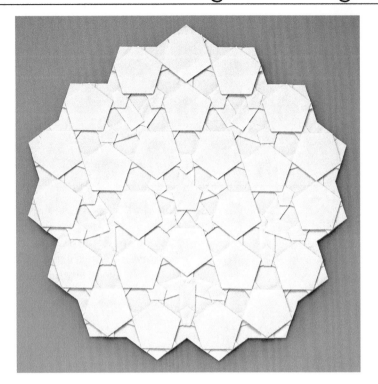

Back side

This work is regular pentagons tiled with a different rule than Figure 4.13. To fill the gaps, the following two tiles were newly designed.

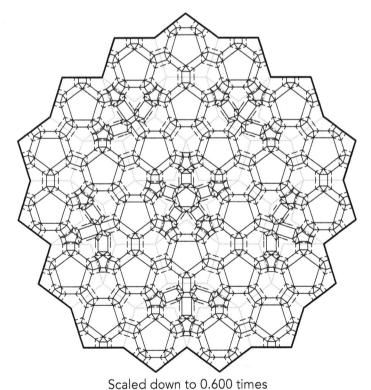

Scaled down to 0.600 times

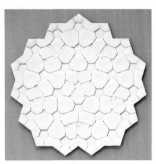

Star tile

Bird foot tile

Artwork 24. Islamic Tessellation #1

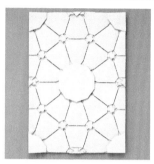

Back side

This is based on the Islamic tessellation in Figure 0.17a, Artwork 13 and the following two tiles are used.

Scaled down to 0.667 times

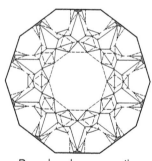

Regular decagon tile

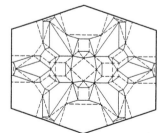

Hexagon tile

Artwork 25. Islamic Tessellation #2

Back side

This is based on the Islamic tessellation in Figure 0.17b, Artwork 15 and the following tile are used.

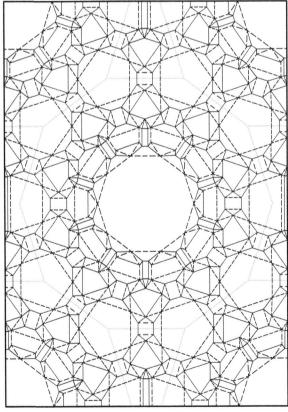

Scaled down to 0.600 times

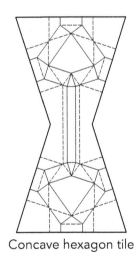

Concave hexagon tile

Artwork 26. Islamic Tessellation #3

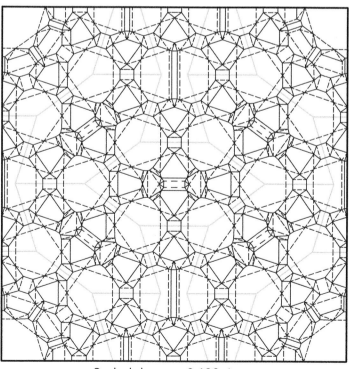

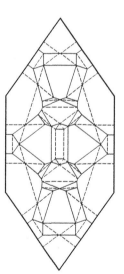

Back side

This is based on the Islamic tessellation in Figure 0.17c. Three types of tiles are used: the two used in Artwork 25 and the following one.

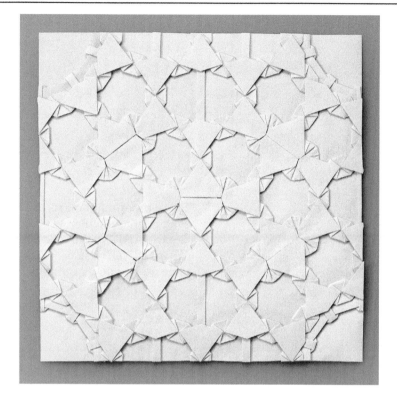

Scaled down to 0.600 times

Hexagon tile

Artwork 27. Penrose Tiling

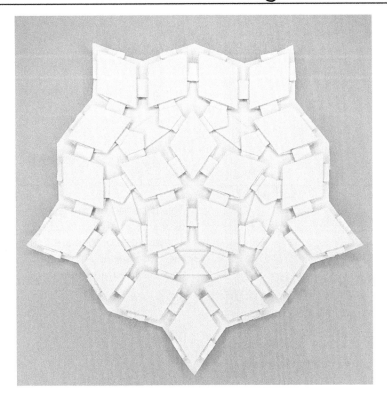

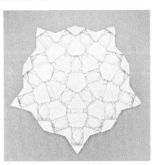

Back side

This is based on the Penrose tiling in Figure 0.18a. Figure 4.15b and the following tile are used. These can be tiled to create aperiodic patterns.

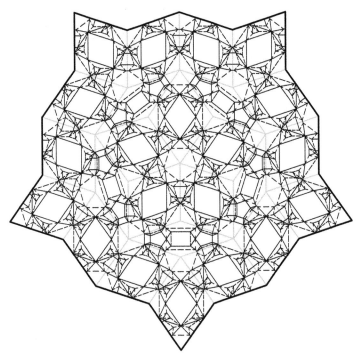

Scaled down to 0.600 times

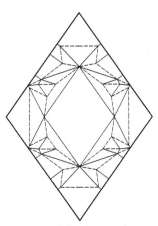

One of the base tiles

Artwork 28. Tiled Snowflakes #1 and #2

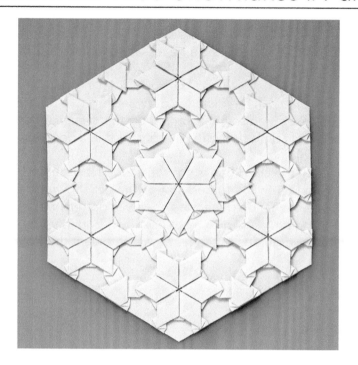

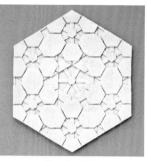

Back side

This is tiled with two types of snowflakes: Artwork 13 and 17. They could not be directly connected, and it was necessary to design the following tile to be used for relaying.

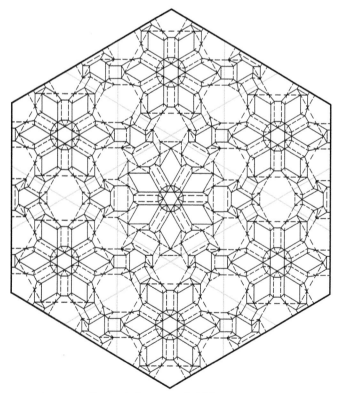

Scaled down to 0.600 times

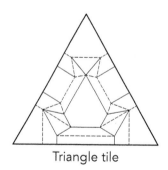

Triangle tile

Appendix 3: Condition that Boundaries are Folded into Similar Shape

This appendix introduces a condition satisfied by pleat bases on a Guide boundary when a scaled-down boundary is formed by flat folding. In other words, it is a necessary condition for forming a scaled-down boundary.

Figure 4.17a shows a scaled-down boundary by flat folding. The scaled-down boundary can be obtained by flipping the pleat bases. We now focus on the distance between any two points chosen from the boundary's vertices and the pleat base's endpoints. Comparing this distance before and after folding, the folded one is less or equal to the unfolded one. This property is a necessary condition for forming scaled-down boundaries by folding.

Figure 4.17b shows an example Guide where the condition is not satisfied. The distance between the vertices, indicated by the white dots, is longer in the folded state. (The arc shown by the thick gray line indicates the original distance.) Since the Guide generates one twist pattern with a rot face intersecting the boundary, it is certainly not possible to form a scaled-down boundary by folding. Even if the interior is considered a set of multiple guides, rot faces always intersect the boundary.

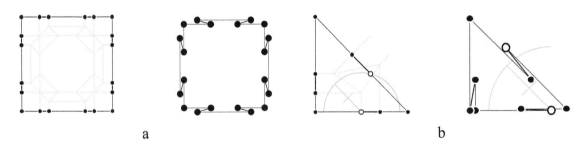

a   b

FIGURE 4.17 Folded boundary

5

Generating Esthetic Origami Tessellations

So far, we have shown the method to design origami tessellations by focusing on the connection of crease patterns. Next, as an application, this chapter shows how to design a set of faces that can be obtained by flat folding. As a result, we can create more esthetically pleasing works. In fact, many of the existing examples have also been created by using the method.

5.1. Origami Tessellations Regarded as Positive–Negative Patterns

Many of the origami tessellations in this book represent patterns that combine black–white polygonal tiles, as described in Section 0.6. These monochrome patterns are called positive–negative patterns (or figure-ground patterns). Origami tessellations can represent positive–negative patterns as a set of faces corresponding to the positive polygons and located above other faces in the folded state. For example, Figure 5.1a can be regarded as the 2 × 2 checker pattern in Figure 5.1b. Also, the works in Figure 5.2a can be regarded as the positive–negative patterns in Figure 5.2b, respectively.

The design of origami tessellations presented in this chapter is that of faces corresponding to positive polygons by using the twist pattern design method. To create intended positive–negative patterns, the design of faces considers their shape and position in the folded state. The designed fold lines do not distinguish between mountain and valley as before, but their positions can be obtained, as described in Section 0.4. Then, by assigning a mountain to each side of these faces, the faces can be located above other faces in the folded state. Note that the crease pattern may not be flat foldable due to MV assignment to other fold lines, and the pattern should be revised.

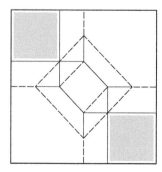
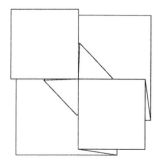
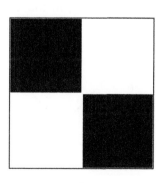

a
b

FIGURE 5.1 Origami tessellations regarded as a 2 × 2 checker pattern

DOI: 10.1201/9781003376705-6

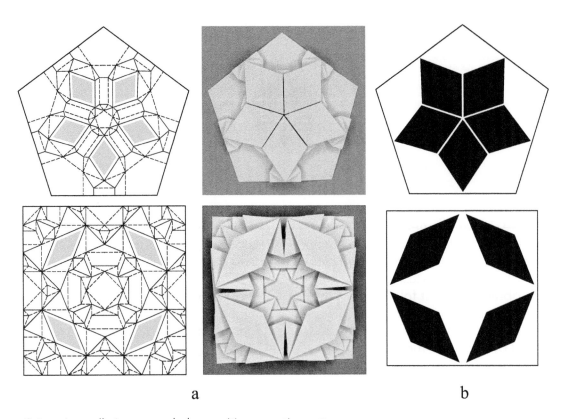

a   b

FIGURE 5.2 Origami tessellations regarded as positive–negative patterns

5.2. Parallel Moving Faces by Flat Folding

First, the relationship between a crease pattern designed by the twist pattern design method and the position of the faces in the folded state is introduced. Comparing the shapes before and after folding, we can see that some faces have moved in parallel without being rotated or flipped. For example, in the case of a crease pattern in Figure 5.3, we can see that the gray faces have moved in parallel. These faces are bounded by different pleats. Since the faces are reflected twice by each parallel fold line contained in the pleats, they consequently can be interpreted as moving in parallel by flat folding.

Next, consider the design of such parallel moving faces. Focusing on a guide vertex, pleats corresponding to the guide side incident with the vertex can bind a parallel moving face. That

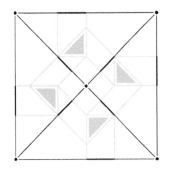
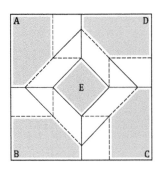
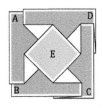

FIGURE 5.3 Parallel moving faces by flat folding

is, a parallel moving face is generated for each guide vertex. Each side of its face is perpendicular to each side incident to the corresponding guide vertex. Knowing these properties, we can formulate a policy to design a guide to generate desired parallel moving faces.

5.3. Design for Origami Tessellations Regarded as Positive–Negative Patterns

Now, we show a design method to create origami tessellations regarding positive–negative patterns. For example, the pattern with a positive pentagonal polygon is shown in

Figure 5.4a, and the work representing it is shown in Figure 5.4b. An important point of the method is that a parallel moving face introduced in Section 5.2 represents a positive polygon.

Given the desired positive–negative pattern and its boundary as in Figure 5.4a, for each positive polygon, determine the elements of the Guide using the following procedure. However, each positive polygon is limited to convex polygons.

First, determine a guide vertex corresponding to the positive polygon and guide sides incident with the vertex. The guide vertices should be placed inside a corresponding positive polygon. Each guide side should be perpendicular to each side of the polygon, but they need not necessarily intersect. For example, Figure 5.4a can yield Figure 5.5a.

Next, determine the ratio of the length of the pleat base to the length of a corresponding side.

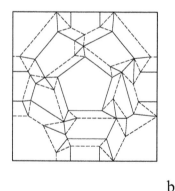

a b

FIGURE 5.4 Origami tessellation regarded as a regular pentagon

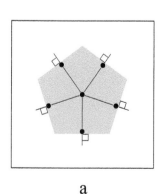
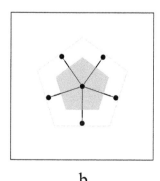
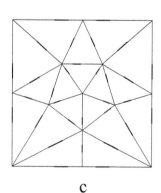

a b c

FIGURE 5.5 Method to design Guide of origami tessellations regarded as positive–negative pattern

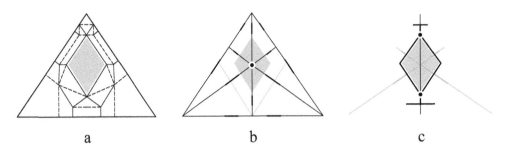

FIGURE 5.6 Guide with sides that don't correspond to a side of a positive polygon

If the ratio is x, then the length of one side of the boundary is 1–2x times by flat folding. Then, scale down the desired positive polygon using the corresponding guide vertex as the center so that each side is 1–2x times. That is, the scale of the polygon is adjusted to the folded state. Place each pleat base so that the pleats are placed along the boundary of the scaled-down polygon. For example, when the ration of the pleat base is determined to 0.2, a positive polygon in Figure 5.5a is scaled down so that each side is 0.6 times its original length, resulting in Figure 5.5b.

Determine the other elements of the Guide to include the elements obtained here. If there is more than one positive polygon, the same method is used to obtain Guide elements for each polygon. Note that the boundary of the Guide coincides with the given positive–negative pattern. Curiously, when a crease pattern generated from the Guides is folded, each face corresponding to the positive polygon moves so that it locates in its original position. We invite you to try out this wonder.

Finally, we show how guide sides that do not correspond to sides of a positive polygon can be incident with the guide vertex corresponding to the polygon. Look at the gray face in Figure 5.6a. The face is generated by a guide in Figure 5.6b to represent a positive rhombic polygon. Six sides are incident with the guide vertex corresponding to the polygon. Two of them have a pleat base to generate a zero-length fold line since they do not correspond to a side of the polygon (Figure 5.6c). Since faces can have zero-length sides, the desired polygon can still be formed in such cases.

Artwork 29. Heart

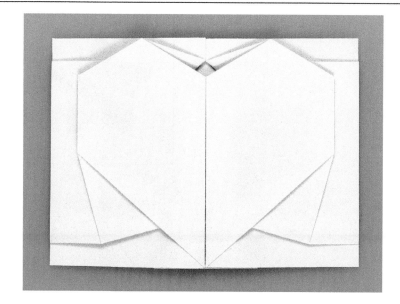

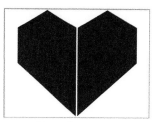

Based pattern

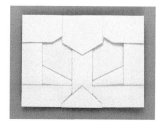

Back side

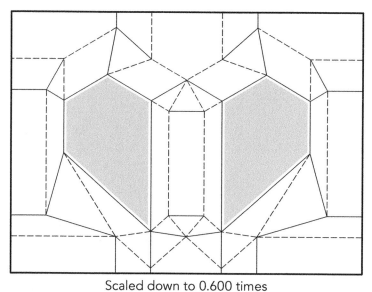

Scaled down to 0.600 times

Guide

This work expresses one heart by using two faces. The boundary is rectangular to minimize the number of fold lines as much as possible.

Artwork 30. Star inside Square

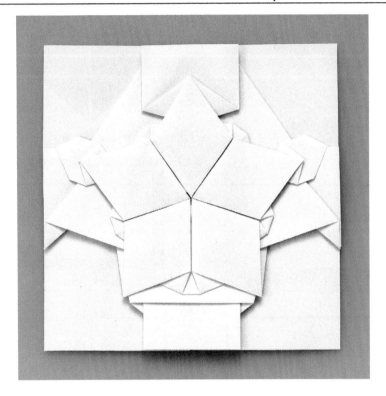

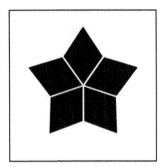

Based pattern

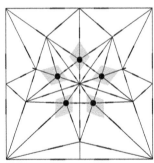

Back side

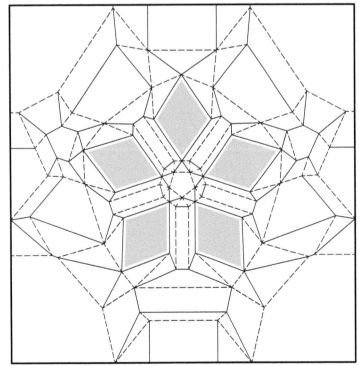

Scaled down to 0.600 times

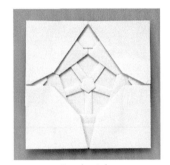

Guide

This work is designed with the square boundary of Figure 4.12 as a constraint. Therefore, it not only expresses a star but can be connected to other patterns. Origami enthusiasts will appreciate the difficulty of designing this work.

Artwork 31. Ohio Star

Based pattern

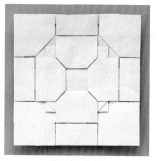

Back side

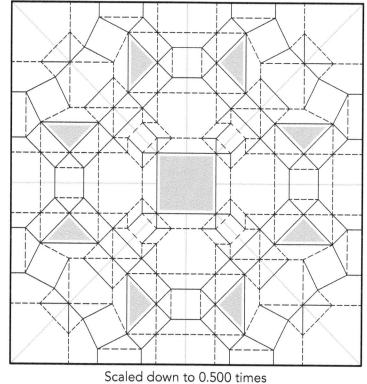

Scaled down to 0.500 times

Guide for based tile

This is based on the Ohio Star, which is used in patchwork. Because the based pattern is rotational and mirror symmetrical, the entire crease pattern can be created by designing only a part, as before.

Artwork 32. Kagome

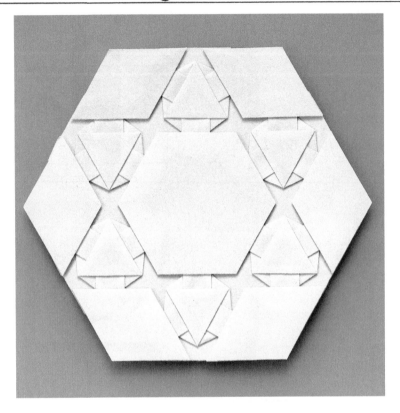

Based pattern

Back side

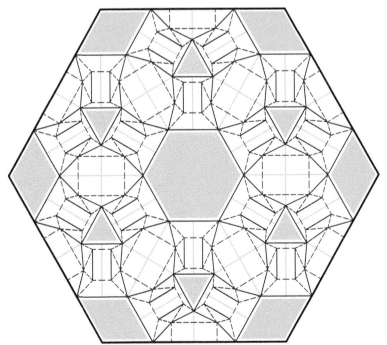

Scaled down to 0.600 times

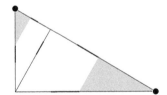

Guide for based tile

The twist pattern design method also excels in inserting gaps of the intended width between positive polygons.

Artwork 33. Asanoha

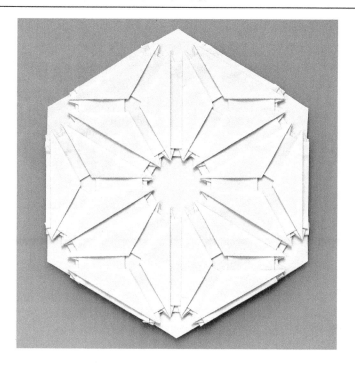

Based pattern

Back side

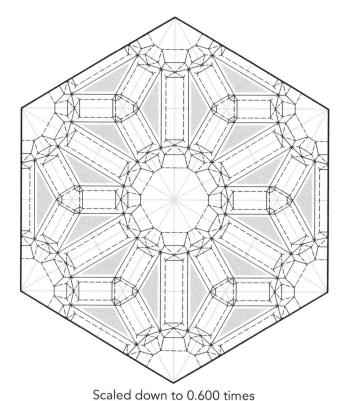

Scaled down to 0.600 times

Guide for based tile

This work required a very elaborate arrangement of the fold lines, as 12 positive polygons were gathered in the center. By repeatedly adjusting the position of the pleat bases, a relatively easy-to-fold pattern was obtained.

Artwork 34. Bishamon-Kikkō

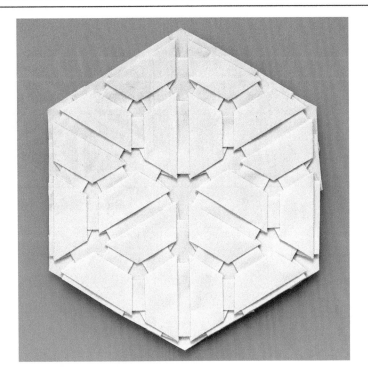

Based pattern

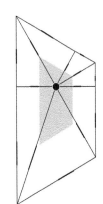

Back side

Guide for based tile

This work has the shape of a regular hexagon obtained by tiling trapezoid patterns. It was difficult to design the Guide for the based tile because the shape was different from what we had done before.

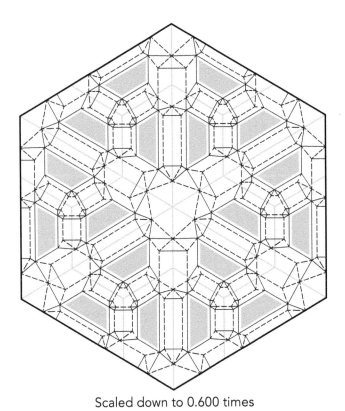

Scaled down to 0.600 times

Artwork 35. Cherry Blossom

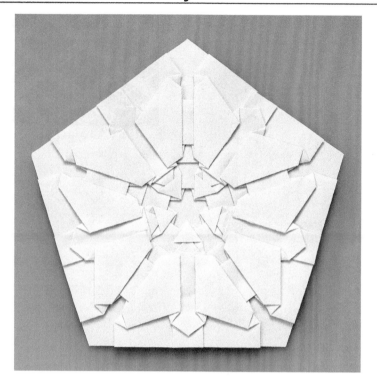

Based pattern

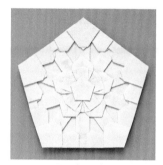

Back side

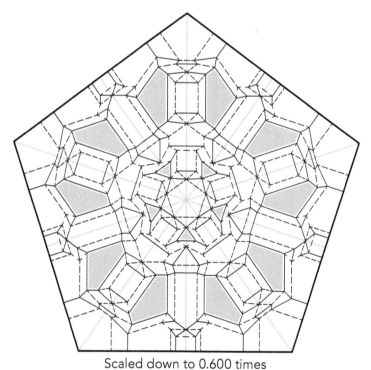

Scaled down to 0.600 times

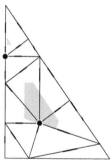

Guide for based tile

There is no fixed positive–negative pattern for cherry blossoms. Therefore, we personally designed the pattern, taking into consideration the ease of Guide design and final folding.

Artwork 36. Takeda-Bishi

Based pattern

Back side

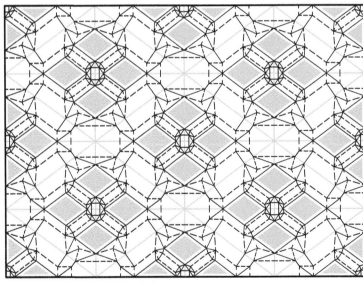

Scaled down to 0.600 times

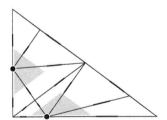

Guide for based tile

This work is based on a pattern including rhombuses. By alternating the design of the based pattern and the Guide, a crease pattern with fewer fold lines was obtained.

Artwork 37. Connected Asanoha and Bishamon-Kikkō

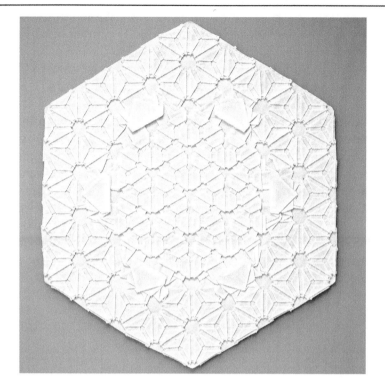

Based pattern

Back side

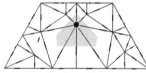

Guide for relay tile

This work is a tiled version of Artwork 33 and 34 via relay tiles. The relay tile is also constrained as a positive and negative pattern, which makes the work cohesive.

Scaled down to 0.600 times

89

Appendix 4: Deformation of Crease Pattern Using Pleat Bases

Crease patterns created with the Triangle Twist Pattern Maker can be deformed into another one by changing the lengths of the pleat bases, as described in Appendix 2. This column introduces the property of such deformation. Using the relation introduced in Appendix 2, interesting deformations can be obtained.

First, Figure 5.7 shows an example of such a deformation applied to a single twist pattern. As the ratio of the lengths of the guide side to the length of the pleat base approaches 1 from 0, the rot face rotates and scales up to coincide with the guide face. When the ratio is 1, the pleat base and guide side coincide. This property is also held when the pattern consists of multiple twist patterns.

The most interesting property of this deformation is that crease patterns are structurally identical before and after changing (Figure 5.8a). For example, the number of lines does not change, and the fold lines with zero length are always zero length. This property is called graph isomorphism. The reason for this can be seen by focusing on faces that moved in parallel without being rotated or flipped by flat folding. As explained in Section 5.2, these faces are bounded by pleats corresponding to guide side incident with a vertex. Let a, b, c, etc. be the distances from the vertex to the pleat bases in that order, as shown in Figure 5.8b. Before and after changing, these a:b:c, etc. ratios are the same. This means that the face always changes only in scale. Other fold lines are constructed by connecting the vertices of different faces. Therefore, the graph isomorphism of the crease pattern is held.

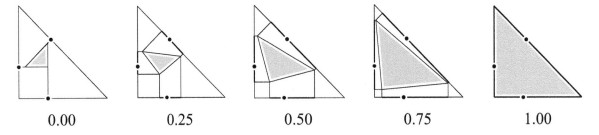

| 0.00 | 0.25 | 0.50 | 0.75 | 1.00 |

FIGURE 5.7 Deformed triangle twist pattern (the value is the length of the pleat base relative to the guide side)

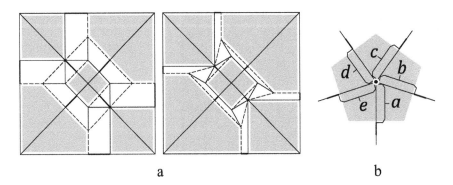

a   b

FIGURE 5.8 Deformed checker base

6

Folding Bellows

The origami tessellations introduced so far have been flat folded. This chapter shows a different approach and tries to create three-dimensional origami works. These works are periodic, based on the bellows. Since their crease patterns are periodic, they can be considered origami tessellations.

6.1. Folding Parallel Lines

First, let's create zigzag patterns by folding parallel lines (Figure 6.0). The folding may sound simple, but it is surprisingly profound. Prepare a piece of paper and make parallel folds to divide into eight equal parts (Figure 6.1a). The first fold line is a mountain fold, the next is a valley fold, and so on, alternating the direction of the folds. The resulting work is stable, with each fold slightly open (Figure 6.1b). This type of origami with a zigzag cross-sectional shape is called "bellows." The number of parallel fold lines of bellows patterns can easily be increased. Bellows with many fold lines can be

beautifully shaded depending on the way the light hits them.

In the bellows as shown in Figure 6.1, the MV assignment and the fold angle can be changed individually for each fold line. These operations can deform the bellows. Figure 6.2 shows deformed bellows in Figure 6.1 so that the fold lines repeat "mountain, mountain, valley, valley" in the order from the bottom. Each fold line forms a right angle, and the cross-section resembles continuous squares. Figure 6.3 shows deformed bellows so that the fold lines repeat "no-fold, mountain, valley" in order from the bottom. Since the heights of the faces are different, the folded state forms sloping faces. Thus, by simply changing the mountains and valleys (or no-folding), new patterns can be obtained, so please give it a try.

6.2. Bent Bellows

Next, let's bend the bellows. To do this, first prepare flat-folded bellows. Fold the center of

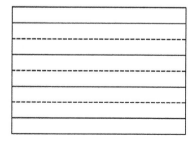

a

b

FIGURE 6.1 Simple bellows

DOI: 10.1201/9781003376705-7

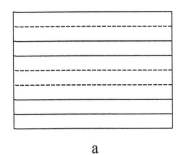

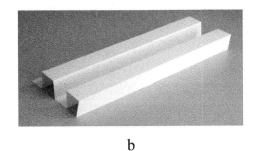

a   b

FIGURE 6.2 Square bellows

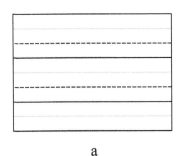

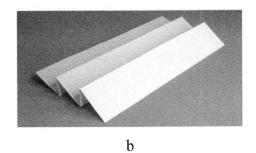

a   b

FIGURE 6.3 Sloping bellows

the bellows as in Figure 6.4a. This state shows zigzag fold lines are added to the crease pattern, as in Figure 6.4b.

The addition of the fold lines creates 4-degree interior vertices in the crease pattern. The set of fold lines near the interior vertex looks like a bird foot. Figure 6.5a shows two types of bird foot in the crease pattern. These can also be flat folded with different MV assignments, as shown in Figure 6.5b. Based on them,

let's change the MV assignment in Figure 6.4 so that the zigzag fold lines are mountain folds. As a result, a crease pattern and the folded state in Figure 6.6 can be obtained. Its folded state looks like bent bellows.

By bending the bellows many times in this way, we can create different bellows. For example, Figure 6.7 shows twice-bent bellows. Try to deform it into various shapes by changing the folding angle, etc.

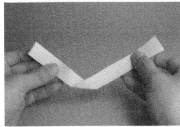

a   b

FIGURE 6.4 Adding fold lines to bellows

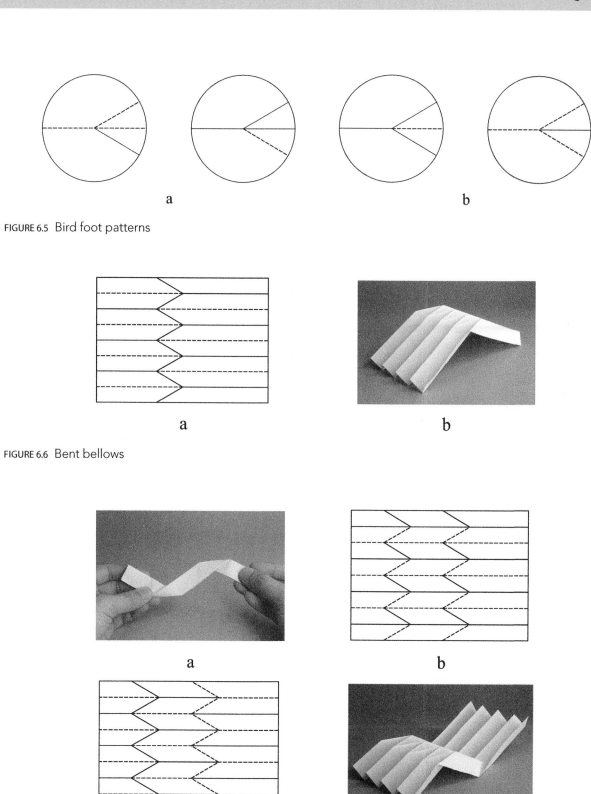

FIGURE 6.5 Bird foot patterns

FIGURE 6.6 Bent bellows

FIGURE 6.7 Twice-bent bellows

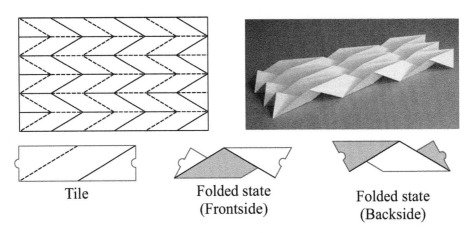

Tile   Folded state
(Frontside)   Folded state
(Backside)

FIGURE 6.8 Periodic bellows

6.3. Periodic Bellows

The crease patterns of bellows introduced in Section 6.2 were vertically periodic. In contrast, we now show a method to design crease patterns of bellows that are horizontally periodic, as shown in Figure 6.8. The basic method is the same as before: create a basic tile and connect them. The crease pattern is created by using a tile in Figure 6.8. The horizontally connected tiles are periodic. Tiles are also connected to satisfy mirror symmetry in the vertical direction, and fold lines are added along the upper and lower boundary lines.

Next, we show that the tile, as in Figure 6.8, can be made from a rectangular piece of paper. First, fold the rectangular paper an even number of times. Figure 6.9 shows examples. If the bottom side of the folded state and the top side of the backside are fully visible, the part can be connected (or the reverse of this). Figure 6.9a,b satisfies this condition.

Figure 6.10 shows examples of connecting the tiles in Figure 6.9a,b. As mentioned above, fold lines are added along the upper and lower sides of the tile. If interior 6-degree (or more) vertices are created in the crease pattern, as in Figure 6.10b, consider that zero-length fold lines

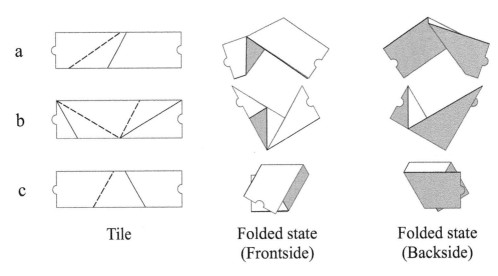

a

b

c

Tile   Folded state
(Frontside)   Folded state
(Backside)

FIGURE 6.9 Folded rectangular papers as tiles of bellows

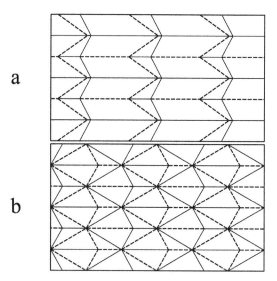

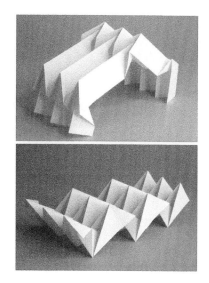

FIGURE 6.10 Periodic bellows using tile of Figure 6.9a,b

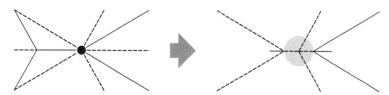

FIGURE 6.11 Set of fold lines around interior vertex that can be regarded as set of bird feet with zero-length fold lines

are hidden, as shown in Figure 6.11. In this way, the set of fold lines around the interior vertex can be regarded as a set of bird feet. According to Figure 6.5b, an MV assignment makes them flat foldable.

6.4. Bending Irregular Bellows

In Section 6.1, we create irregular bellows by changing the MV assignment. For these bellows, we can also bend as in Section 6.2.

First, focus on the sloping bellows as shown in Figure 6.3. As before, a fold is added to the flat-folded state shown in Figure 6.12a, and then bird feet is added to the crease pattern in Figure 6.12b. MV assignments are changed as shown in Figure 6.12c. By folding the pattern again, bent bellows in Figure 6.12d can be formed.

On the other hand, square bellows, as shown in Figure 6.2, cannot be flat folded, so the approach to bend must be changed as follows. First, prepare a crease pattern of bent bellows, as shown in Figure 6.7. Divide this crease pattern along the horizontal fold lines. Each divided strip is slid vertically to create a gap. For example, Figure 6.13a shows four strips obtained from the pattern in Figure 6.7c. The fold lines used for the division are maintained on the boundaries of the strip. Then, fold lines are added by connecting the same vertices as before the division (Figure 6.13b). The connected fold lines are assigned the same mountain or valley fold as the upper and lower zigzag fold lines. By folding this crease pattern, we can obtain a bent square bellow as shown in Figure 6.13c.

Of course, these ideas can also be applied to the periodic bellows described in Section 6.3. Try combining various techniques to create new works.

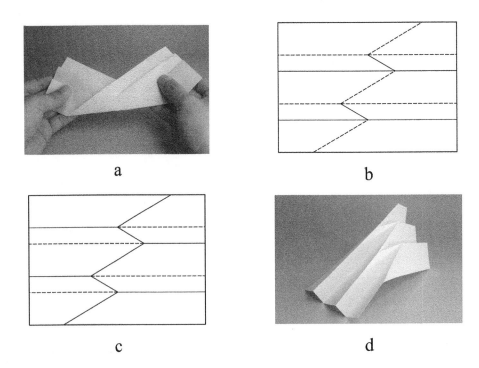

FIGURE 6.12 Bent sloping bellows

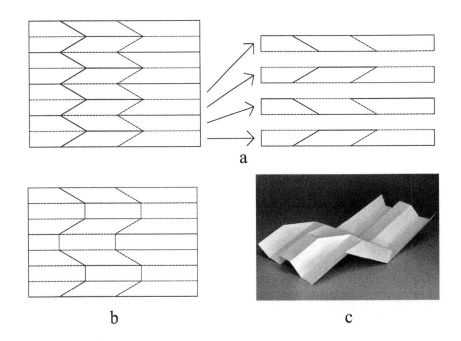

FIGURE 6.13 Twice-bent square bellows

Artwork 38. Continuous Trapezoids

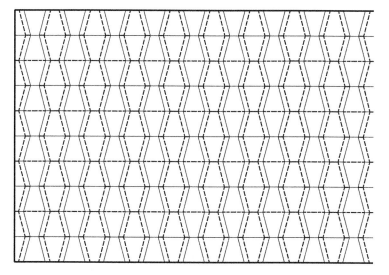

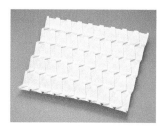

Back side

Even simple bellows with continuous trapezoidal faces can produce beautiful shadows when illuminated.

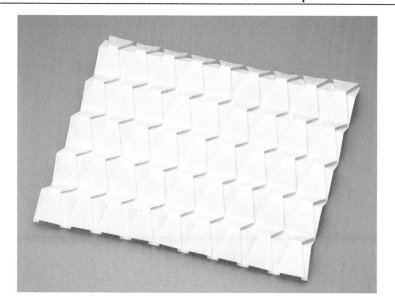

Artwork 39. Mountain Range

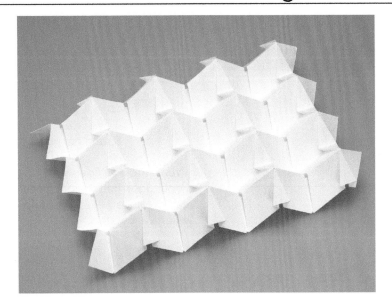

Back side

This forms a zigzag shape like a mountain range. The shape is simpler because some faces are hidden.

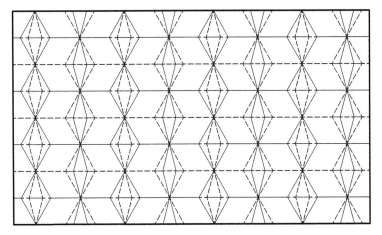

Artwork 40. Block Bellows

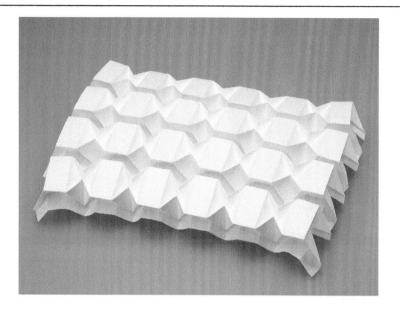

Back side

When these bellows are shrunk, the shape is more solid than expected. Therefore, this is one of the works that we hope you will actually make.

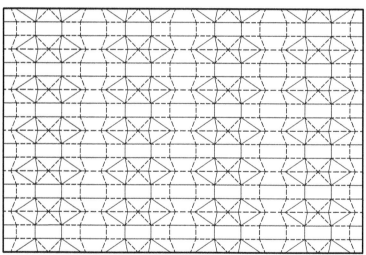

Artwork 41. Domed Bellows

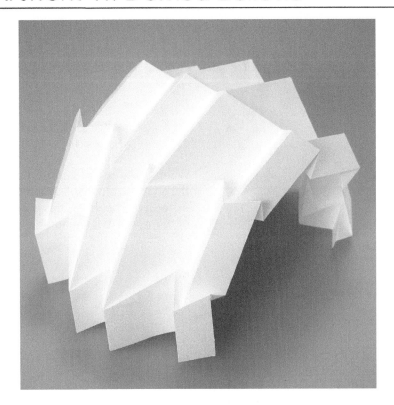

Back side

This work is made by folding sloping bellows in an arc. As a result, it stabilized in a hemispherical dome-like shape.

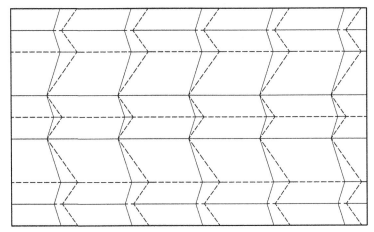

Artwork 42. Hilbert Curve

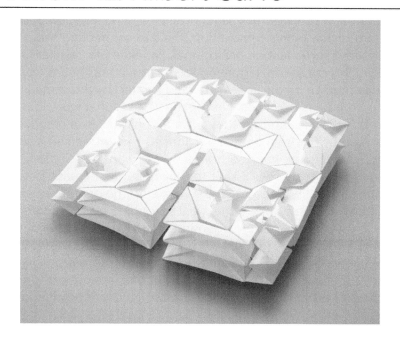

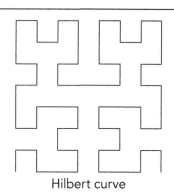

Hilbert curve

The bellows represent a Hilbert curve, one of the self-similarity structures. Since the crease pattern is very long, we present it divided into four pieces.

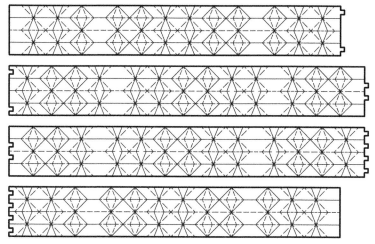

Column 03: Origami Tessellation Design Software "Tess"

The software Tess, developed by Alex Bateman, is widely known for generating crease patterns with connected twist patterns[1]. The generation procedure is as follows. First, a tessellation and corresponding center points for each tile are given. In this tessellation, the line connecting the center points of adjacent tiles and the side that they share must be orthogonal. For example, regular tessellation and semi-regular tessellation satisfy this condition. Figure 6.14a shows a 2 × 2 square tessellation as an example. Next, each tile is scaled down by the same ratio and rotated by the same angle. Then, the sides of the deformed tile are treated as fold lines. Finally, the vertices that were identical before the deformation are connected by fold lines so that they form a polygon. As a result, twist patterns are created with the face enclosed by the connected fold lines as the rot face. In a similar way, the crease pattern shown in Figure 6.14b can be created from a 4 × 4square tessellation.

Compared to the twist pattern design method, there is a limit to the variation of crease patterns that can be created with Tess. For example, all the twist patterns that make up the generated crease pattern with Tess have the same angle between the side and the pleat extending from the side. The twist pattern design method can theoretically create patterns generated by Tess. However, if each twist pattern of the tessellations is generated from one guide, the Guide isn't periodic, even though the crease pattern is periodic. The bold line in Figure 6.14c is the boundary of the periodicity unit representing a single twist pattern. Since the pleat and the side of the boundary are not orthogonal, no guides coincide with these boundaries. Therefore, it is necessary to use a no-periodic Guide, as

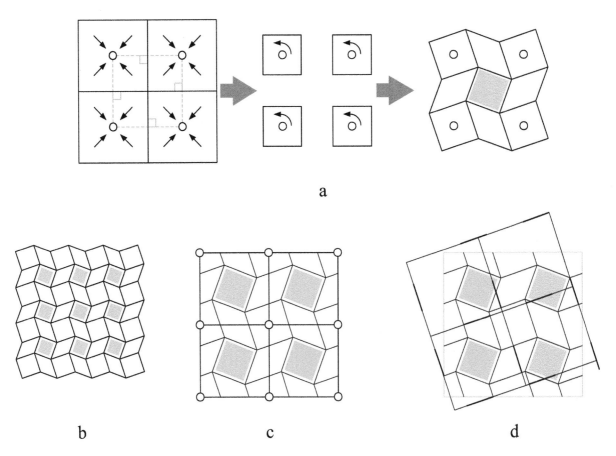

a

b   c   d

FIGURE 6.14 Crease pattern designed by Tess

shown in Figure 6.15d. When creating patterns, we should choose a design method that is suitable for it.

Note

1 "Tess: Origami tessellation software." http://www.papermosaics.co.uk/software.html (accessed on April 26, 2023).

7

Application of Twist Pattern Design Method

We have shown how to design origami tessellations using the twist pattern design method in stages. Now you should be able to create a wide variety of works. However, for those who want to create more original works, this chapter shows some tips for applying the twist pattern design method.

7.1. Reconstructing Guide from Given Origami Tessellation

In Chapter 3, it was explained that a crease pattern consisting of triangle twist patterns could be constructed by using the twist pattern design method. The opposite is also true, i.e., we can reconstruct a Guide from a given pattern consisting of twist patterns. This section shows how to reconstruct them through several case studies.

First, reconstruct a guide from a single triangle twist pattern, as shown in Figure 7.1a. Since the rot face and the guide face are similar, make such a guide face from the given rot face. The scale can be determined freely, but each guide side should be orthogonal to the corresponding pleat. The guide is reconstructed by repositioning each pleat base over each guide side (or its extension), as shown in Figure 7.1b. There is more than one type of guide to be reconstructed. For example, the rot face that is located outside the guide face can also be reconstructed, as shown in Figure 7.1c.

Guides can also be reconstructed from crease patterns consisting of multiple twist patterns. In other words, a set of reconstructed guides for each twist pattern is the Guide for a given pattern. The scale and position of each guide should be determined as follows.

- The scale should be determined so that the ratio of the lengths of the sides to the

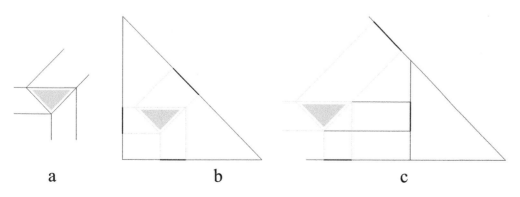

a　　　　　　　　　b　　　　　　　　　c

FIGURE 7.1 Reconstructing Guide for a single triangle twist pattern

DOI: 10.1201/9781003376705-8

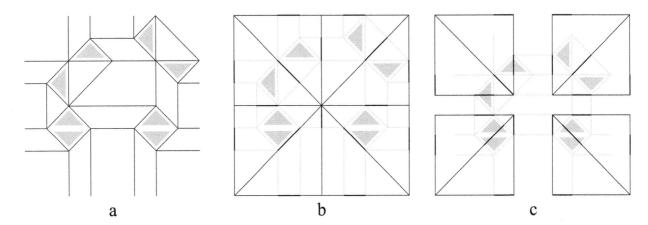

a   b   c

FIGURE 7.2 Reconstructing Guide for crease patterns consisting of multiple twist patterns

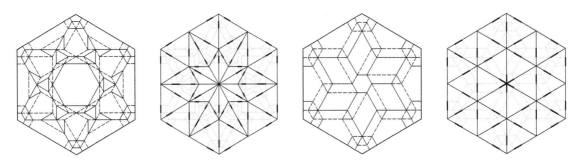

FIGURE 7.3 Negative space star and star twist

lengths of the pleat base is the same in all guides.

- The position should be determined so that the guides of twist patterns sharing a pleat share the side.

When a rot face is not a triangle, it should be divided into triangles and interpreted as a set of triangle twist patterns, as described in Section 5.2. For example, from a crease pattern shown in Figure 7.2a, a Guide shown in Figure 7.2b can be reconstructed.

If the position condition is ignored, multiple Guides with gaps can be recreated, as in Figure 7.2c. In this case, we can see that the guide sides facing each other across a gap are parallel and that pleat bases corresponding to each side create an identical pleat. In other words, gaps are interpreted as rectangle-shaped areas inserted between the guides.

Examples of applications of this feature are presented in Section 7.3.

For example, Figure 7.3 shows Guides reconstructed from works presented in *Origami Tessellations* (Eric Gjerde, 2008), which is a textbook on origami tessellations. The ability to reconstruct Guides will allow the various origami tessellations to be handled in a uniform manner.

7.2. Fractal Origami Tessellations and Guides

Fujimoto's *Hydrangea*, a well-known pattern, has a feature that a part of the object resembles the whole. For example, by adding lines to the crease pattern shown in Figure 7.4a, the crease

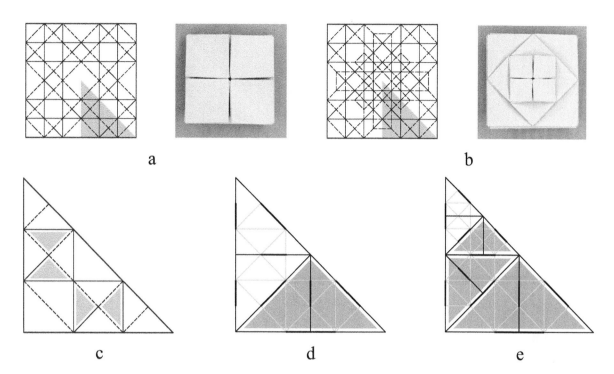

FIGURE 7.4 Reconstructed Guides from *Hydrangea*

pattern shown in Figure 7.4b is created, which has the same pattern as shown in Figure 7.4a in its center. This feature is referred to as fractal. As a tip for creating fractal origami tessellations, here are fractal Guides reconstructed from these crease patterns. The triangular crease pattern shown in Figure 7.4c is the part indicated by the gray triangle shown in Figure 7.4a. This crease pattern can be interpreted as four connected triangle twist patterns, although some overlapped fold lines are removed. Therefore, the Guide shown in Figure 7.4d can be reconstructed. Similarly, the Guide shown in Figure 7.4e can be reconstructed from Figure 7.4b. Interpreting the part of Guide, indicated by the gray triangle, as the unit, we can see the whole is constructed by placing the scaled-down units toward the center. (The white Guide corresponding to the center is also identical, but the MV assignment to fold lines is different.) Thus, by placing scaled-down units, fractal crease patterns can be created. Examples of works are shown at the end of this chapter.

7.3. Guide with Gaps

In Guide reconstruction introduced in Section 7.1, we recommended that adjacent guides share sides. However, there are cases in which it is easier to understand the structure of Guides if the position condition is ignored. In this section, we introduce such Guides with gaps.

Figure 7.5a shows the checker base introduced in Figure 1.11. Since it consists of four triangle twist patterns, a Guide can be reconstructed. As introduced in Section 1.3, the crease pattern shown in Figure 7.5 can be created by connecting 16 checker bases. A Guide in Figure 7.5c is reconstructed by replacing each checker base in Figure 7.5b with Guide in Figure 7.5a. By allowing for a Guide with gaps, a Guide for crease pattern in Figure 7.5b can be interpreted as a concatenation of identical checker-based guides.

Finally, we show a method for creating such a Guide with rectangular gaps. The Guides can be created by deforming tessellations in

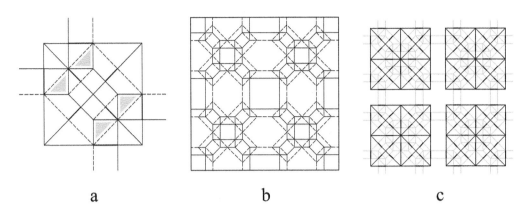

FIGURE 7.5 Guides reconstructed from checker base and crease pattern by connecting checker bases

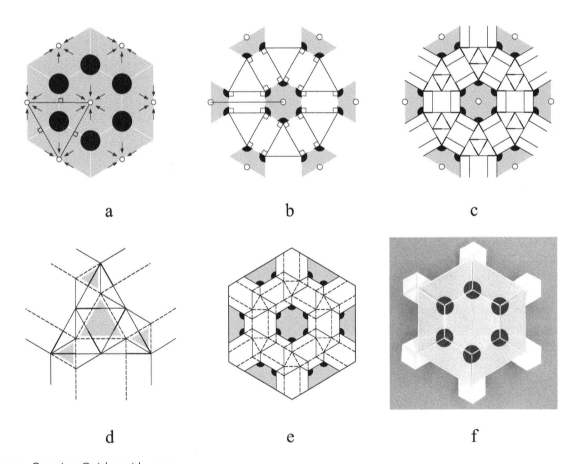

FIGURE 7.6 Creating Guides with gaps

which the line connecting the center points of adjacent tiles and the side that they share are orthogonal. Such a tessellation condition is called spider web condition. Tessellation satisfying the condition is often used for designing

origami tessellations, such as Tess introduced in Column 03. Figure 7.6a shows a tessellation that satisfies the condition. Each tile is scaled down by the same ratio, and polygons are created by connecting the vertices that were in the same place before the scaling (Figure 7.6b). There is a rectangular area (gap) between the polygons. By placing Guides whose boundary coincides with the polygons, a whole Guide with gaps can be created (Figure 7.6c). For example, from the Guide shown in Figure 7.6c, a crease pattern shown in Figure 7.6e is generated by concatenating the patterns in Figure 7.6d.

Using Guides with gaps, scaled-down tiles (areas shown in gray) can be included as faces in the crease pattern. Since each side of the tile is tangent to a rectangle gap and is parallel to the pleats through the gap, the faces as tiles can be created by adding the fold lines of the pleats along each side. In particular, when a crease pattern from Guides with the ratio of the pleat base to the sides as 0.5 is created, a set of tiles of the folded state reconstructs the original tessellation. For example, Figure 7.6f shows the folded state of Figure 7.6e and represents Figure 7.6a. Thus, Guides with gaps can be used to represent the original tessellations as origami tessellations. Examples of works are shown at the end of this chapter. On the other hand, note that the boundary is not scaled down by folding since the boundary of the paper does not coincide with the boundary of the Guides.

Artwork 43. Deformed Hydrangea

Back side

This work is a fractal pattern introduced in Figure 7.4, created by shortening the length of the pleat base. The outside petals are created by adding folds to the pleats extending outward from the base tile.

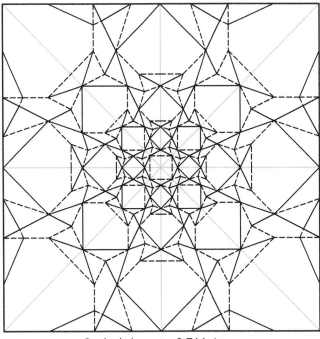

Scaled down to 0.714 times

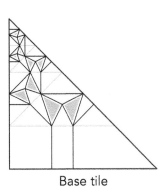

Base tile

Guide for base tile

Artwork 44. Radial Checker

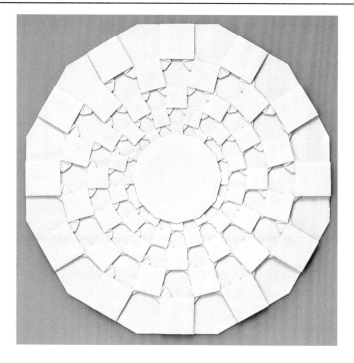

Back side

This work is also a fractal pattern made by applying the technique learned from the hydrangea to the checker base. The outer square faces are created by extending the pleats outside the Guide.

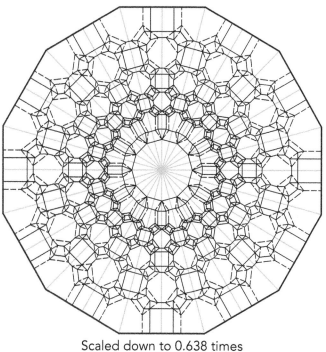

Scaled down to 0.638 times

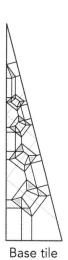

Base tile

Guide for base tile

Artwork 45. Sangi Kuzushi

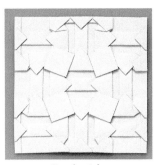

Back side

This work appears to be simple tiled rectangles but is realized by intricate fold lines. By slightly changing the pattern on the inside and outside, the rectangular faces can be placed near the boundary as well.

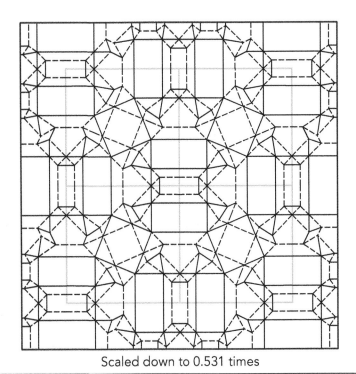

Scaled down to 0.531 times

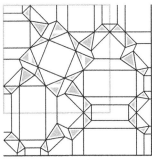

Pattern in the lower right neighborhood

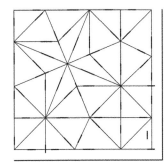

Guide in the lower right neighborhood

Artwork 46. Kazaguruma (Windmill)

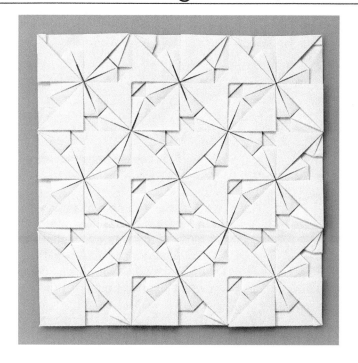

Back side

This work, like Artwork 45, has a different inner and outer pattern. The internal pattern is made from the same set of guide faces as in Artwork 45. So, the two works are very similar.

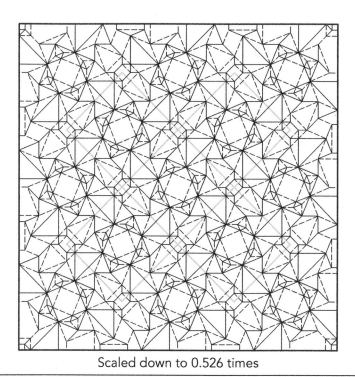

Scaled down to 0.526 times

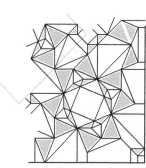

Pattern in the lower right neighborhood

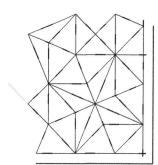

Guide in the lower right neighborhood

Artwork 47. Uroko (Scale) #3

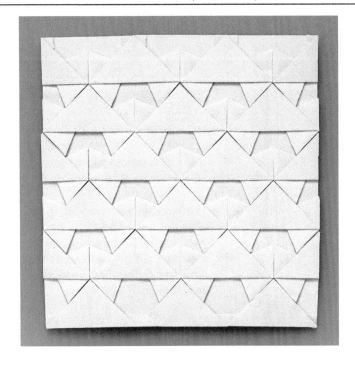

Back side

This is an example of a work generated by a Guide with gaps. In this way, by creating Guides from positive–negative tiling scaled down, we can create works expressing the tiling.

The short sides are scaled down to 0.666 times

Base tiling

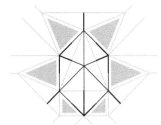

One of the Guides

Artwork 48. Yamaji (Mountain Pass)

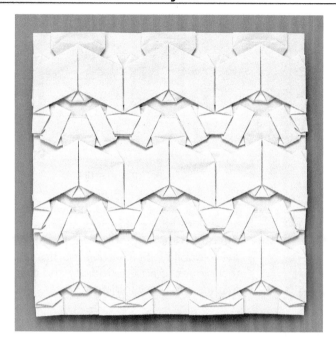

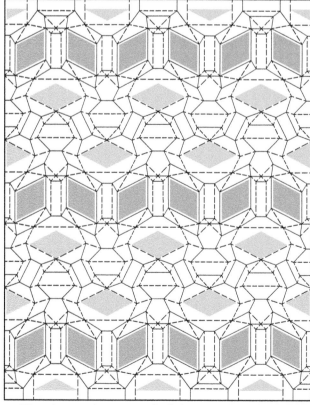

The short sides are scaled down to 0.600 times

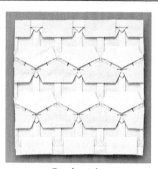

Back side

Depending on the base positive–negative tiling, several Guides need to be designed. The Guide presented here corresponds to the area near the vertices shown in black in the base tiling.

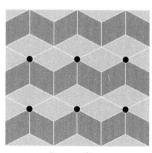

Base tiling

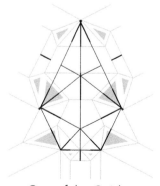

One of the Guides

Artwork 49. Yagasuri (Arrow Feathers)

Back side

This work is one of those that went through several iterations of revisions. Since the proposed method only guarantees flat foldability, it still requires artisanal techniques to add periodicity and to create fold lines for easy folding.

Scaled down to 0.500 times

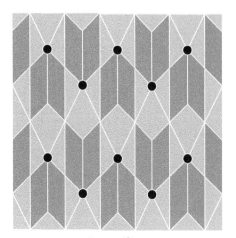

Base tiling

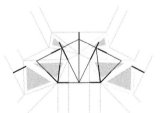

One of the Guides

Artwork 50. Music Note

Back side

The positive–negative tiling on which this work is based is called a Voronoi diagram. This complex work can be created because a set of tiles of a Voronoi diagram can represent an arbitrary polygon.

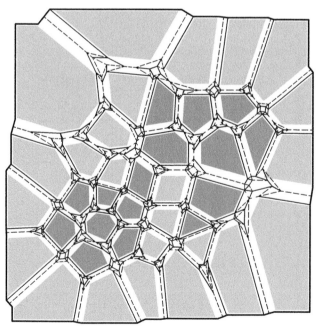

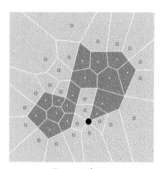

Base tiling

One of the Guides

Column 4: Connecting 3D Origami Arts and Origami Tessellations

The crease patterns introduced so far, in which the boundary is scaled down by folding, are not limited to being flat folded. For example, the "whipped cream" in the book *3D Origami Art* (CRC, 2016), shown in Figure 7.7, is of a three-dimensional shape, but its boundary is scaled down by folding. Therefore, it can be connected with the works introduced in this book, as shown in Figure 3.14.

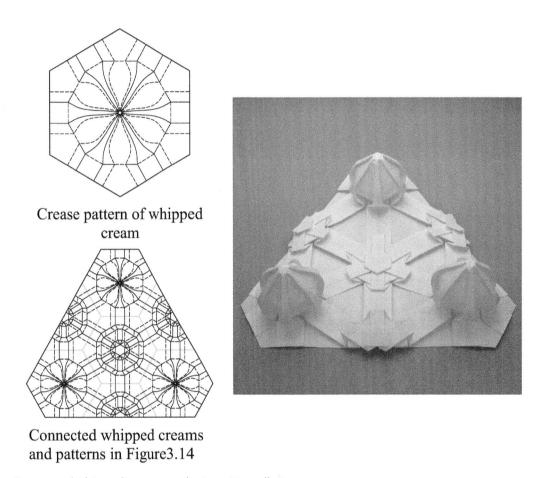

Crease pattern of whipped
cream

Connected whipped creams
and patterns in Figure3.14

FIGURE 7.7 Connected whipped creams and origami tessellations

References

Toshikazu Kawasaki, *Bara to Origami to Suugaku to* [in Japanese], Morikita Publishing, 1998.

Erik D. Demaine, Joseph O'Rourke, *Geometric Folding Algorithms: Linkages, Origami, Polyhedra*, Cambridge University Press, 2007.

Eric Gjerde, *Origami Tessellations : Awe-Inspiring Geometric Designs*, A K Peters/CRC Press, 2008.

Jun Mitani, *3D Origami Art*, A K Peters/CRC Press, 2016.

Robert J. Lang, *Twists, Tilings and Tessellations : Mathematical Methods for Geometric Origami*, A K Peters/CRC Press, 2018.

Jun Mitani, *Curved-Folding Origami Design*, A K Peters/CRC Press, 2019.

Thomas C. Hull, *Origametry: Mathematical Methods in Paper Folding*, Cambridge University Press, 2020.

Robert Fathauer, *Tessellations: Mathematics, Art, and Recreation*, A K Peters/CRC Press, 2020.

Tomoko Fuse, David Brill, *Tomoko Fuse's Origami Art: Works by a Modern Master*, Tuttle Publishing, 2020.

Index